A CRASH COURSE

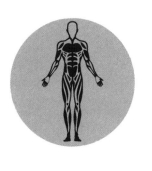

ANATOMY

Joanna Matthan

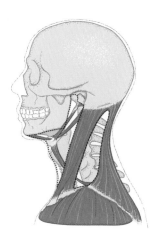

IVY PRESS

First published in the UK in 2019 by
Ivy Press
An imprint of The Quarto Group
The Old Brewery, 6 Blundell Street
London N7 9BH, United Kingdom
T (0)20 7700 6700 **F** (0)20 7700 8066
www.QuartoKnows.com

© 2019 Quarto Publishing plc

British Library Cataloguing-in-Publication Data
A catalogue record for this book is available
from the British Library

ISBN: 978-1-78240-859-8

This book was conceived, designed, and produced by
Ivy Press
58 West Street, Brighton BN1 2RA, United Kingdom

Publisher Susan Kelly
Art Director James Lawrence
Editorial Director Tom Kitch
Project Editor Elizabeth Clinton
Design JC Lanaway
Illustrator Robert Brandt
Design Manager Anna Stevens
Series Concept Design Michael Whitehead

Printed in China

10 9 8 7 6 5 4 3 2 1

A CRASH COURSE

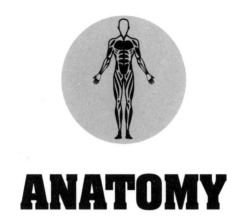

ANATOMY

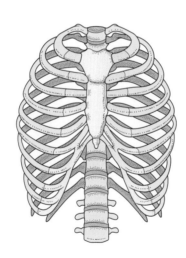

INTRODUCTION

The human body is a wondrous structure and learning how it works is a constant source of amazement to the inquisitive mind. Not only does knowledge of the body and how it functions equip us with a wealth of "fun facts" with which to impress others, it is perhaps the best way to understand how our life choices affect our own physical well-being, helping us to make sense of who we are and what happens within us when things go wrong. Broadly-speaking, most structures in our body have a known purpose and exist to allow us to function to our maximum potential. There are organs without which we simply cannot live. Life is impossible, for instance, without the brain as a control center to regulate the way our bodies work or without a beating heart to pump blood and oxygen around the body. Modern medicine (surgery in particular) has made it possible to replace or remove some of what traditionally was considered essential to life. It is now possible to live without a lung or kidney or spleen or part of a liver. A dialysis machine can clean our blood in a similar manner to the kidneys. Our hearts can continue beating with pacemakers, or even with a completely new heart from an organ donor. We can replace limbs with prosthetics. In theory, today it is possible to live without arms, legs, eyes, ears, nose, breasts and many of our internal organs, and have a rich life. Still, a wholesome body allows us to experience and interact with the world around us in many different ways that enhance each other and add layers of detail that we may only properly notice when things go wrong. Anatomy is the study of the structure of the body (and its component parts), through which we can learn about the functioning of our body and develop a new appreciation for its complexities.

The anatomical landscape

The landscape of the human body has been charted meticulously through untold hours of grueling—and often smelly—work by doctors and anatomists of old dissecting decomposing bodies in intolerable conditions. They cut up (anatomy means "to cut up") and mapped out the topographical atlas of our body, making their way through unknown territories and recording their findings in painstaking detail, not unlike the voyagers who once traversed the wide and unexplored oceans. Without these pioneers, we would have remained in the dark ages of medical knowledge, our understanding of the human body shrouded in mystery and misunderstanding and subject to superstition. And yet this landscape, despite its relative familiarity to those in the healthcare professions, still remains elusive and somewhat mysterious to many of us, even today.

Through ancient and medieval times, knowledge of the human body remained the secret realm of the few physicians and healers who dedicated themselves to curing ailments— and, in ancient Egypt, to those who embalmed the bodies of the dead to aid them on their journey in the afterlife. Physicians from ancient times until the 15th century CE observed and treated patients based on an imperfect understanding of what lay under the skin, restricted by law and superstition to looking only skin deep. Animals were dissected initially to understand how the body worked, and only rarely did physicians catch glimpses of internal organs or observe the interplay of structures beneath the skin. Later, when dissection became more commonplace, it remained within the confines of medical establishments and linked to hospitals. Doctor-cum-anatomists dissected to understand the workings of the body during a period when textbooks and illustrated anatomy atlases were still unavailable. Even though public dissection for a time caught the interest of the general public (and, sometimes, the criminal cunning of body-snatching entrepreneurs), it did so in a somewhat grotesque manner, with the shock value of seeing a dead body cut up arguably of more interest to many than actually understanding the intricacies of the human body, particularly as the bodies of criminals were publicly humiliated in this manner. Today, anatomical knowledge is available everywhere, and exhibitions parading plastinated and dissected bodies draw in millions of visitors around the world, revealing closely guarded secrets to the general public and making the body more accessible for learning.

Anatomy in modern times

The true foundation of medicine (and, by extension, all healthcare professions), anatomy is the most concrete of all the sciences healthcare students will encounter in their learning journey. Dissecting the human body used to be a requirement for all medical students in previous times. Today, financial constraints, lack of qualified staff, limited access to donated bodies, and the requirement for complex and cumbersome equipment to embalm and preserve them (as well as stringent regulations around body donations) mean that numerous institutions around the world have cut down foundational anatomy teaching for students. Moreover, where it does happen, instead of full-body dissections, other tools are increasingly utilized, with technology now a significant factor in how the subject is taught.

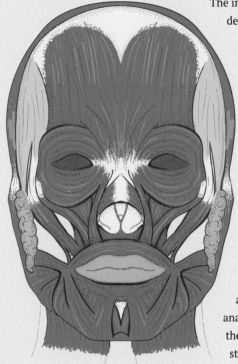

The internet is rife with anatomy video clips and tutorials detailing the secrets of the human body for anyone keen to learn. Countless software applications for smartphones and computers are available to further enhance their knowledge. Digital and paper anatomy textbooks, exquisitely illustrated atlases, lifelike plastic models, virtual reality glasses, and all manner of additional equipment are also competing for attention in anatomy curricula.

Still, the general public and healthcare students (as well as numerous clinicians and anatomists) remain firmly convinced that there is no replacement for learning anatomy from donated bodies. Among other things, none of the other approaches allow the student to learn and appreciate the differences in the human body due to age and sex—or to experience the range of common anatomical variants they are likely to encounter in the course of their clinical careers. Instead of requiring students to dissect, many institutions now produce pre-dissected material available for learning. These prosections, as they are called, may be embalmed or plastinated (longer-lasting). Surgeons may train on fresh-frozen bodies donated specifically for this purpose, as these are more realistic for learning procedures. Access to donated bodies to dissect or learn anatomy from through prosections is still a valuable experience for students, not solely for the tactile feedback they get from direct handling of body tissues and perceptions of depth that can only really be appreciated with a hands-on approach, but as a rite of passage and a formative experience in learning how to be professional and sensitive when faced with death and dying, the bread and butter of most healthcare professions.

Our anatomy is the same but different

On the surface we all look different, unique even, although we do have certain features in common. People tend to think that below the surface we are all anatomically fairly uniform. This is untrue. Not only are there differences between men and women and the different age groups, there may be considerable differences between two individuals of the same sex and age. Anatomical and medical textbooks often depict the most common arrangement of our anatomy, the average person. But when describing the human body, the word "normal" represents a wide range of variations, all of which are compatible with life and do not diminish our ability to function fully.

In anatomy, variations are those manifestations of our similar anatomy that, for whatever reason, have followed another path but arrived at the same (or similar) destination. It may mean some of us have a few muscles that others do not have; for instance, *risorius* in the face, *palmaris longus* at the wrist, *psoas minor* at the back of the abdominal wall, or *pyramidalis* at the front of the abdominal wall. We may have a different network of blood vessels supplying certain organs—or our appendix may measure anywhere between ¾ and 8 in. (2 and 20 cm) and project in a variety of directions from its point of origin on the large bowel. Even more curious, while on the outside we look mostly symmetrical, several of our internal organs are unpaired and therefore asymmetrical. During fetal development, a signaling mechanism exists to "induce" this asymmetry, leading to the normal organization (*situs solitus*) of the organs in our chest and abdomen, with the heart, stomach, and spleen on the left side and the liver and gallbladder on the right. A most curious and extremely rare (1 in 10,000) anatomical variation is when the large internal organs are mirrored or reversed from their more common positions. In *situs inversus*, all of the internal organs are mirrored, so the heart, stomach and spleen are on the right side and the liver and gallbladder, amongst other organs, found on the left. Even rarer is *dextrocardia*, when just the heart is swiveled around so its apex, rather than pointing to the left, points to the right. Both of these variations are compatible with a completely normal life, although there may be a higher incidence of heart and other problems among this specific population. We may live our entire lives with a total lack of awareness of the variations that lie beneath the surface, as they rarely have a significant impact on how we function.

Variations are fun to learn about. They illustrate to us how very complex and multifaceted our bodies are and how our bodies can function even with a slightly different collection of structures to the "norm". But, even more importantly, anatomical variations are crucial information for healthcare professionals, and surgeons in particular. Knowledge of these "normal variants" allow us to anticipate the unexpected and avoid mistakes that can be life-threatening if missed.

Getting the most out of this book

The anatomical knowledge available today is extensive. The most up-to-date edition of *Gray's Anatomy*—the main source of inspiration for this book and the modern-day "Anatomy Bible"—has over 1500 pages and weighs over 11 lb. (5 kg), every sentence packed with critical and measured information, representing the phenomenal explosion in anatomical knowledge over the last few decades. Yet, even this compendious tome only scrapes the surface of what we know about the body, so much information is there now available from so many sources.

There is no one perfect way to learn about the human body but, in the absence of a cadaver to dissect, anatomy is possibly best learned through a combination of images and text, and through the ages this has been the preferred method of learning. This book distils anatomical knowledge into manageable chunks, allowing you to choose whether to skim-read or delve a bit deeper into the secrets of our complex flesh-and-bones machines. Anatomical detail has been included on some, but not all, areas of the body, and on some, but not all, structures within it. The Introductory section gives you an overview of some of the key concepts and the language you might need to navigate the sometimes murky waters of "anatomy speak". Some of the many body systems are considered at various points in the book for those wanting more than a quick read. The four main anatomy chapters—**Head & Neck Anatomy, Thoracic Anatomy, Abdominopelvic Anatomy**, and **Back & Limb Anatomy**—have been divided into 13 topics each.

Chapter 1, **Head & Neck Anatomy**, cruises through some of the important structures within the head and neck region, such as the two main groups of muscles for facial expression and mastication, the salivary glands that moisten our food, and some of the sense organs (tongue, ear, and eye). Neuroanatomy and dental anatomy (teeth) are only touched on in this book, as both are complex anatomical subject areas in their own right, each requiring a dedicated book to do them justice.

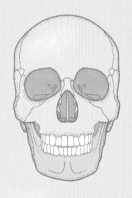 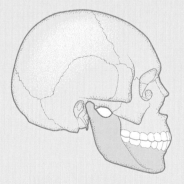 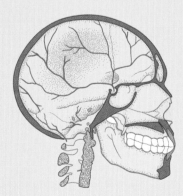

Chapter 2, **Thoracic Anatomy**, highlights significant areas within the chest: the heart and lungs and their coverings, the esophagus, and the lymph-carrying thoracic duct. The breast is included within this section, despite belonging anatomically under the skin system; as part of the external covering of the chest, it gives a more rounded appreciation of the area.

Chapter 3, **Abdominopelvic Anatomy**, covers a wide area from the diaphragm below and showcases some of the more prominent structures in the abdominal, pelvic and perineal (genital) regions.

Chapter 4, **Back & Limb Anatomy**, travels through the upper and lower limbs and the back, all of which enable movement. Like a guided city tour, only some of the body's main sites are featured. At times, off-beat and obscure locations are pointed out to spark greater interest that may lead you to a fuller investigation of this ever-fascinating subject.

How to use this book

This book distills the current body of knowledge into 52 manageable chunks, allowing you to choose whether to skim-read or delve in a bit deeper. There are four chapters, each containing 13 topics. The introduction to each chapter gives an overview of various systems in the body.

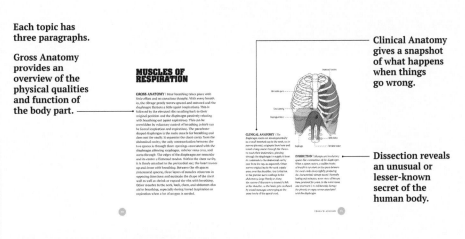

Each topic has three paragraphs.

Gross Anatomy provides an overview of the physical qualities and function of the body part.

Clinical Anatomy gives a snapshot of what happens when things go wrong.

Dissection reveals an unusual or lesser-known secret of the human body.

Body systems and anatomy know-how are covered on pages 16-27, preceded by a timeline of anatomical milestones and biographies of key influencers.

TIMELINE

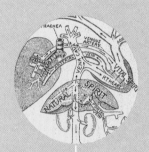

HUMAN VIVISECTION

The Ancient Greeks methodologically dissected animals to understand the workings of the human body. Herophilus and Erasistratus were Alexandrian surgeons who also "laid open men" that were still alive (usually criminals), basing their scientific observations on these poor unfortunates. This, human vivisection, was justified as having benefits for decent people.

Ancient world

3rd century BCE

2nd century BCE

ANATOMY IN THE VERY ANCIENT WORLD

The Ancient Egyptians paved the way for anatomical studies. A surgical manual from 1600 BCE has case studies on trauma; several internal organs were recognized but not all of their function was understood. Greek medical theorist and natural philosopher, Alcmaeon of Croton (5th century BCE), observed that reason was seated in the brain, not the heart. He was an advocate for anatomical dissection.

GALEN'S INFLUENCE AND ERRORS

The Greek physician Galen increased knowledge of anatomy through treating wounded gladiators, but also dissected pigs and apes. He performed successful cataract surgery and proved that urine is formed in the kidney, not the bladder. But he also decided that blood ebbed and flowed from the heart, an erroneous belief that persisted until the 1600s.

REBIRTH OF ANATOMICAL STUDY

Leonardo da Vinci dissected and sketched the eye and the optic nerves, the brain and a fetus *in utero*. Andreas Vesalius wrote *De humani corporis fabrica* (On the Structure of the Human Body) challenging accepted anatomical knowledge. Inaccuracies in Galen's erroneous theories were exposed. William Harvey accurately described the circulation of blood.

ANATOMY IN MODERN TIMES

Anatomy dissections and access to cadavers became closely regulated after the Anatomy Act of 1832. In recent times, radiological techniques have allowed connections to be made between anatomy and physiology. Study of anatomy has also been integrated with other disciplines (genetics, biochemistry, biophysics, biomechanics). PET, CAT, and MRI scanning enable an internal view of the body without performing surgery or even dissection.

15th century CE

17th–19th century

Modern times

ANATOMY THEATERS AND PUBLIC DISSECTIONS

Anatomical theaters arose around the world and dissection was performed in these theaters for anyone to observe. Public dissections were performed on hanged bodies. The physician Henry Gray wrote *Anatomy Descriptive and Surgical*, currently known as *Gray's Anatomy*, which has now run to over 40 editions and is still the definitive book on the subject today.

BIOGRAPHIES

HEROPHILUS (335–280 BCE)

Known as the father of anatomy, Herophilus was born in the ancient maritime town of Chalcedon, on the shore to the east of the Bosphorus Strait, in the eastern part of what is now modern Istanbul. He was one of the first to do public dissections on human cadavers (sometimes on live criminals, too) during a short period in Greek medical history when human dissection was allowed. Herophilus wrote meticulous accounts on his dissections, and his study of the brain led him to believe it (not the heart) was the center of the intellect. He wrote about the cavities (ventricles) within the brain and traced the blood-filled spaces (sinuses) within the tough outer covering of the brain (dura mater). He also distinguished nerves from blood vessels and tendons, and was the first to measure arterial pulse, now so commonly used in clinical practice. Herophilus wrote a commentary on Hippocrates, a handbook for midwives, and works on the causes of sudden death and on anatomy. All of his work was lost when the library of Alexandria was destroyed in 272 CE, although he is quoted by several later anatomists. He died in 280 BCE.

GALEN (130–210 CE)

A Greek physician, writer, and philosopher, Galen was born in Pergamon in Asia Minor, modern-day Bergama in Turkey. He was *the* major influence on medical practice and theory in Europe, the Middle East, and the Byzantine world through the Middle Ages until the 17th century. As the son of a rich architect, he was educated to be a philosopher, but changed to a career in medicine at the age of 16, which he studied in his hometown as well as at Smyrna on the coast and Alexandria in Egypt. Galen came to be known as a physician–philosopher who served four Roman emperors, the most outstanding among all anatomists, and the "Prince of Physicians". Above all he was a methodological anatomist who extensively documented his work. Galen gleaned his initial anatomical knowledge from treating wounded gladiators. He chose to dissect pigs (human dissection was banned) to learn more about internal organs (pigs are anatomically somewhat similar to humans), but several of his conclusions were wrong. These wrong beliefs persisted in medical circles for a millennium, at times impeding progress altogether as they came to be considered infallible. Galen wrote over 300 works, many of which are lost or are texts that are attributed to him via inclusion in other works.

MONDINO DE LUZZI (1270–1326)

Known as the "restorer of anatomy," Mondino was an Italian physician and anatomist born in the city of Bologna, Italy, the medieval center of medical learning. His rejuvenation of the systematic study of anatomy had far-reaching consequences for medical and surgical knowledge. De Luzzi's father and grandfather were pharmacists, but he trained to be a physician at the University of Bologna and practiced both medicine and surgery after that, all the while studying and teaching anatomy. Systematic teaching of the subject had been abandoned for centuries until De Luzzi reintroduced it to the curriculum. Moreover, the ban against human dissections had been lifted in Medieval Europe. Such operations were often carried out by underlings, but De Luzzi, preferred to carry out this task himself, performing public dissections and giving lectures as he did so. Based on the teachings (some erroneous) of Galen and the Greek and Arab anatomists of old, De Luzzi's handbook of anatomy—*Anathomia Mundini* or *Anathomia corporis humani*—started a new era of spreading anatomical knowledge. The book was written in 1316 and first printed in 1478. The first dissecting manual produced, it went through 39 editions and was used for 250 years. De Luzzi died in 1326 having left a far-reaching legacy.

LEONARDO DA VINCI (1452–1519)

Better known to the world as an artistic genius and polymath, Leonardo made, arguably, his greatest contribution in the field of human anatomy. Born in the town of Vinci, near Florence in Italy, Leonardo was apprenticed to the famous Florentine artist Verrocchio at the age of 14, who expected his pupils to make a deep study of anatomy. Later, he dissected more than 30 cadavers; this first-hand experience and his remarkable ability to reproduce his observations both verbally and in beautiful sketches contributed greatly to developing anatomical knowledge. He was the first to describe a cirrhotic liver and to accurately draw the human spine in a manner that explores its biomechanics. His drawing of a fetus within the womb is the first scientific drawing of its kind. Against the prevailing view that the heart was a two-chambered structure, Leonardo observed it had four chambers, and was cone-shaped but with a rotational twist. He also performed experiments with an ox's heart to understand how the aortic valve worked. Leonardo died in 1519, at the age of 67, following a series of recurrent strokes. Through his skilled dissections and wondrous ability to draw, he left an unsurpassed legacy of over 240 detailed anatomical drawings and 13,000 words on the subject.

THE LANGUAGE
OF ANATOMY

Navigating the complex terrain of the human body requires use of a language that is understandable across borders and disciplines: international and interprofessional. Thousands of anglicized anatomical terms originate from ancient Latin and Greek and are still used worldwide to avoid confusion. This adds richness to anatomy as a discipline where, today, knowledge of the meanings of the original words, when explained to students, deepens their understanding of the concepts. Knowing what the original word meant often helps us appreciate the shape of a structure or remember its function—for instance, "pelvis" means basin or bowl. Eponymous terms are those named after the person thought to have discovered them. Anatomists are divided on their use; many feel they confuse learners. Clinicians sometimes take a very different view, so for the moment it may be safer to teach both. For instance, anatomists might use "uterine tubes" to describe the two tubes that project off the uterus. Clinicians may prefer the eponymous "Fallopian tubes," after Gabriele Falloppio who first described them, although several use the two terms interchangeably. To complicate matters, the tubes resemble a trumpet-like wind instrument with a wide end—known as a salpinx—and so an infection of the uterine/Fallopian tubes is *salpingitis* (*-itis* indicates infection), far removed from the original anatomical term (uterine). Terms like the circle of Willis, ligament of Bigelow, foramen of Monro, and the great vein of Galen hint at a pioneering spirit and arduous labor behind sometimes very small structures. To lose these links would be to lose some of the richness and history of the discipline of anatomy.

Navigating through planes and directions

All anatomical descriptions of the body must begin with a common understanding of the terms used to indicate site and location when discussing them. An upright person, standing straight, eyes looking ahead, arms by their sides and palms facing forward, lower limbs together and toes pointing forward is in the anatomical position. In this position, the penis and the tongue need to be upright. With this anatomical position in mind, four imaginary planes cut through the body and divide it into sections, as it is easier to understand the body in two-dimensional slices. These planes also have specific anatomical names that immediately allow you to orientate to the view you need to take. They are the median, sagittal, transverse, and coronal planes. The median plane passes vertically through the body, dividing it into right and left sides. The sagittal plane can refer to this plane, too, if it is in the midline (mid-sagittal) or it can be any vertical plane that is parallel to the median plane (para-median). The coronal (or frontal) plane divides the body into front (anterior) and back (posterior) halves. The transverse (or axial) plane is a horizontal plane at right angles to the median and sagittal planes.

We can navigate through the body accurately using terms that describe the relative position of structures within the body, always based on the anatomical position. Structures close to the trunk are *proximal* and those away from the trunk are distal: the fingers are distal to the elbow, the elbow proximal to the fingers. Structures closer to the midline are medial, and those to the sides are lateral: the cheeks are lateral to the nose, the nose medial to the cheeks. Structures toward the head are superior (or rostral—rostrum means "nose"—or cephalic or cranial) and those lower down are inferior (or caudal—caudal refers to the "tail" and is used when talking about developmental anatomy): the abdomen is superior to the thighs, the knee inferior to the thigh. Structures more to the front are anterior (or ventral) and those to the back are posterior (or dorsal): the buttocks are posteriorly located, the penis anteriorly. Structures are superficial when they are closer to the skin, and deep when they are further away from the skin: the ribcage is superficial to the lungs. Ipsilateral structures are on the same side of the body, contralateral on the opposite side of the body. Structures on both sides are bilateral.

Movements

Movements can happen in three different planes—transverse, coronal, and sagittal—which can also overlap, for instance, when rotating an arm at the shoulder (circumduction). Movement away from the midline is abduction and toward the midline adduction. So, moving your arms away from the side of your body is abduction and bringing them back into the anatomical position is adduction. When the angle of a joint is narrowed, the movement is flexion. When the angle is widened, the movement is extension. Bending your fingers to grasp an object is flexion, straightening them is extension.

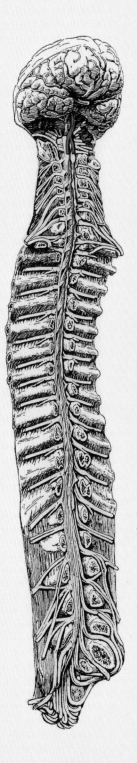

SURFACE ANATOMY

The contours and undulations of the body cover deeper structures, protecting them but also revealing subtle clues about the structures that lie inside. Healthcare professionals learn surface landmarks of the deeper structures to work out what may be going on below the surface, to help describe their findings precisely, and to make accurate diagnoses based on location.

Imaginary lines help locate structures

The anatomical areas at the front of the body are divided up by imaginary lines drawn on the body. Several of these lines can be located by the structures that underlie them, such as bony bumps or muscles. So, knowing where the clavicles are on either side of the sternum (breastbone) helps work out their midpoint, and the vertical line drawn downward from these are the midclavicular lines. The armpit region is the axilla. Its depression reaches all the way to the clavicle, and lines running through the axilla are axillary lines. These lines are useful when placing a chest drain to drain fluid from around the lungs. The pectoral region is the upper chest. The breasts lie here and the pectoral muscles under them. You can feel the lower border of the ribcage (the costal margin). In the midline, the pointy tip of the breastbone is easily felt. An imaginary line running under the ribcage is the subcostal line. An imaginary line (transtubercular line) running horizontally across the upper rim of the pelvis, skimming over the left and right iliac crests, helps locate a safe place to draw fluid out of the vertebral canal. The abdomen is divided into nine different regions by the two midclavicular lines running perpendicular to the subcostal and transtubercular lines. Knowledge of structures within each of these areas can help narrow down and identify potential causes of abdominal pain.

Landmarks in the head and neck

The prominent bump found just behind both of your ears is the mastoid process, the origin of a muscle (sternocleidomastoid) that runs diagonally across the neck to attach on to the clavicle. It divides the neck into triangles in front and behind the muscle, useful for diagnosing neck lumps. Overlying the muscle is the external jugular vein, which when engorged reveals problems relating to the flow of venous blood returning to the heart. You can feel the zygomatic bone and arch (cheekbone) in your face. Above the arch and behind the outer border of the eye socket is the weakest point in the skull (pterion). If you clench your teeth you should be able to feel the temporalis muscle tightening here, and the pulsation of the superficial temporal artery. In the midline of the neck, you should feel a bump halfway down the neck. This is the laryngeal prominence (Adam's apple), a site used to find emergency access into the airway. Low in the midline of the neck, above the collarbone (clavicle) and on either side of the trachea, is the thyroid gland. It moves when you swallow.

Landmarks in the rest of the body

The bony notch in the lower part of the neck is the suprasternal notch. Push backward and you can feel your trachea (windpipe). Run your finger along the sternum (breastbone) until you come to a subtle bump, the manubriosternal angle (or angle of Louis). This is a surface landmark of clinical significance, as moving your finger to either side of the sternum at this point brings you to the second rib, from which it is easy to count down to other ribs. The heartbeat at the tip of the heart (apex beat) can be felt if you work your fingers down to the fifth rib and then place your fingerpads on the space under the fifth rib in the midclavicular line. You can feel a bony bump (external occipital protuberance) in the midline at the back of your head. By running a finger down the midline from this point, you can feel the first bony protrusion at the bottom of the neck, the seventh cervical vertebra (vertebra prominens). This point is used to count down to the lower vertebrae. The blade-shaped shoulder bone (scapula) can be felt pointing downward on both sides. Above the scapula, rising into the neck is a thick muscle on either side, the trapezius. Squeezing tightly on this muscle can help check a person's response to pain, if they are unconscious. The lower back between the thorax above and the pelvis below is the lumbar region. The gluteal region is the area between the iliac crest and the furrow between the thigh and the buttocks (gluteal fold).

ANATOMICAL HIERARCHY

Our bodies are living organisms more complex than anything imaginable. They are capable of doing many different things. Breathing, digesting, thinking, moving, feeling, and creating new life are all seemingly simple and straightforward functions but, really, they require a complex network of multiple systems that constantly communicate with each other, switching tasks and allowing us to do these things without much (or any) conscious contribution. Our bodies are also able to speed up or slow down processes instantaneously, as the need arises. This is all enabled by an organizational hierarchy increasing in size from cells to tissues to organs to systems.

Our shared genetic code is made up of around 3 million bases of deoxyribonucleic acid (DNA), the blueprint for all life-forms, and is present in every cell in our body. DNA is compressed into tightly coiled packages so it can fit into each cell; it works like an instruction manual on how to assemble the thousands of proteins that determine who we are. The adult body is made up of an estimated 75 trillion cells, and millions of these undergo a process of renewal every day. Cells are very small and versatile. The biggest ones measure about the width of a single strand of hair. There are around 200 different types of cells in the body, each adapted to a particular function. Some of them work alone, such as oxygen-carrying blood cells, while others gather together into tissue (muscles, bone, etc.), some of which, when grouped together, form an organ. The organs are then part of a specific body system, and cells with very different functions join together to achieve different tasks within a system.

Seamless interaction

Categorizing body functions helps us understand the specifics of how the body works. Covering us all over is the integumentary system, or the skin (and hair, nails, and breast), our first line of defense against the world of pathogens. Skin regulates temperature and gets rid of waste through sweating. Ligaments, tendons, and cartilage connect the 206-odd bones (and the teeth) that form the supporting framework of our bodies, the skeletal system. It helps us move, produces blood cells, and stores calcium. Layered onto the bones are around 650 muscles, the muscular system. Muscles are specialized by function: those that move us voluntarily are skeletal muscle, those found inside organs and moving internal substances are smooth muscle, and the muscle that pumps blood around the body in the heart is cardiac muscle. The circulatory system, consisting of the heart, blood vessels, and blood, carries blood, nutrients, and hormones around the body, and shifts oxygen and carbon dioxide in and out of cells. The respiratory system, consisting of the trachea, lungs, and diaphragm, allows us to breathe in oxygen and get rid of carbon dioxide. These two systems work together to deliver oxygen to cells.

The digestive system is a pipework of connected organs from mouth to anus that breaks down and absorbs nutrients as well as expels waste. The liver and pancreas contribute to its smooth running. The urinary system consists of the two kidneys, ureters, bladder, and urethra, all of which work together to expel urea from the body. The interplay between the urinary system and the circulatory system controls blood pressure. The immune system protects us against harmful viruses and bacteria, and includes the spleen, bone marrow, white blood cells, and lymph nodes. It works together with the lymphatic system through interconnected lymph ducts and vessels to move the fluid which contains white blood cells and helps fight against infection.

The nervous system includes the central nervous system (brain and spinal cord) and the peripheral nervous system (nerves connecting every other part of the body to the central nervous system). Together they control involuntary actions to sustain life (such as breathing) and voluntary action (such as speech). The nervous system works together with the endocrine system, a collection of eight major glands around the body that secrete hormones vital to life and growth. Our reproductive system allows us to continue the human race. In the male, the testes produce sperm. In the female, the ovaries produce eggs and the fertilized egg grows in the uterus. The reproductive system and the endocrine system work via a complex feedback system based on varying levels of hormone in the bloodstream.

SKIN, NAILS & HAIR

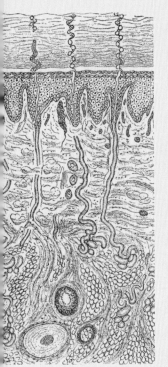

The integumentary system (from the Latin word *integumentum*, meaning "covering") is made up of the skin and structures that develop from it. These are our nails, hair, sweat glands, sebaceous glands, subcutaneous fat and deep fascia, and the breasts. Early signs of illness are often visible on the skin and in the nails, and these are often closely examined by clinicians.

Skin

Our largest organ—skin—envelops our entire body. It dips inside our noses and ears and even covers part of our eardrums and the edges of the conjunctiva in our eyes. Skin forms 8 percent of our total body mass. If you stretched out and weighed the skin of an individual 71 in (1.8 m) tall and 198 lb. (90 kg) heavy, it would spread over an area around 24 sq. ft. (2.2 sq. m) wide and weigh around 15 lb. (7 kg). Our skin self-renews and constantly grows; we shed dead skin cells all the time. Skin has three main layers. The epidermis is the outermost layer, formed of several constantly renewing layers of cells (keratinocytes). The dermis is the inner layer and is composed of dense connective tissue. It contains the blood vessels and nerves that supply the skin. Under this layer is the hypodermis, also known as superficial fascia, a loose layer of connective tissue.

Our skin protects us from the elements but has several other functions essential to life. It can, to an extent, protect us from microbes that would be harmful to us if they entered our bodies, from mechanical and chemical damage, and from ultraviolet rays. Skin forms an interface between the inside of our bodies and the external world. Through its surface, we can sense the world around us via touch, temperature, and texture. Our body temperature is regulated through the skin, which communicates with the nervous and circulatory systems to do this. We sweat when we are overheating, we shiver to warm our bodies. The thermoregulatory center of our bodies in the brain is always trying to maintain a set point, similar to a thermostat. Sweating gets rid of waste products, mainly salts in our body, so it is little wonder that sweat tastes salty. Skin is highly sensitive to touch and also works as a warning system through our ability to feel pain. When we expose our skin to sunlight, a complex cascade of chemical events that involves the liver and kidneys helps produce the vitamin D that is vital for our bones. The activity of melanocytes (specialized cells) that produce the pigment melanin gives us our most marked differences in skin color.

Nails

Nails are self-repairing. They protect our fingertips and the ends of our toes but also make them more sensitive to touch and pressure. Our nails are a compact combination of minerals (such as calcium) arranged into two or three horizontal layers of keratin-filled plates embedded into a protein matrix (a crisscross arrangement). This arrangement gives them their hardness. The main parts of a nail are the nail plate, embedded in the nail folds on three sides of the nail, the nail matrix, and the hyponychium (the free bit at the end of the nail that appears to grow). Growth is constant and happens from the pale semicircular area at the base of the nail (lunula). Fingernails tend to grow faster in young people (and, strangely, in the summer) and grow three to four times faster than toenails. On average, our nails grow around 0.1 millimeters per day. In six months, a fingernail is completely replaced. For a toenail, this happens every 18 months.

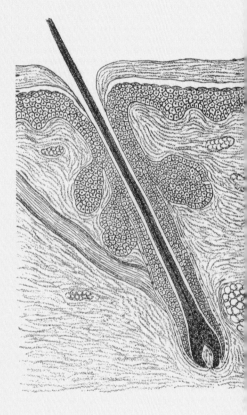

Hair

Any hair you can see on your body is in fact dead. Only the roots of our hair, embedded within the deeper layers of skin, are alive. The amount and variety of hair on our bodies varies greatly and is genetically determined. Hair is found on almost every surface of the body, its thickness and texture varying greatly with where it occurs. On our head and around our genitalia (and beards), hair is thick and coarse. In most other places, hair is fine and soft, and mostly invisible. There are several areas on the body where there is no hair. Notably, on the palms of the hands and soles of the feet there is no hair, nor is there any at the tip of the penis (glans penis) or the clitoris or around the thin skin of the bellybutton (umbilicus) and around the nipple. Hair color is determined by the type and degree of a type of protein (melanin) found in it. As we grow older, our hair loses its melanin and turns gray and white. Individual hair grows in three cycles: at any one time some hair is being shed, some is growing in length, and some is newly emerging. Hair growth is highly variable, though scalp hair grows the fastest. Hair does not grow after death, and cutting hair or shaving a beard do not affect the speed at which they grow.

MUSCLE & BONE

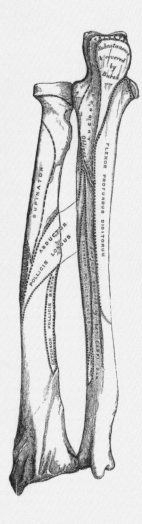

The human skeleton is the framework of around 206 interlinked bones attached to one another via tendons, ligaments, and muscles. Two or more bones meet at joints, which help transmit forces between them, allowing movement or weight to be transferred. Several soft tissue structures support these joints. Ligaments attach bones to one another, so that movement can take place but also to keep movement within safe limits so that the joints remain stable. The skeleton gives our body its shape, bearing the weight of all the tissues that overlie it. Without the skeletal framework and its uniquely human features, we would not be able to stand up straight or move around on two feet. During fetal development, most of our bone is cartilage (both bone and cartilage are connective tissue). This transforms into bone throughout infancy and childhood. Bone is constantly being remodeled, with around 10 percent of an adult's skeleton being remodeled annually.

The bones of the skeleton protect the vital organs within. The skull and facial skeleton are formed of 22 bones surrounding the delicate brain, protecting it from minor bumps to the head and giving our faces their form. Six small bones are found in the ears, enabling hearing. Seven cervical (neck) vertebrae uphold the head and support the neck, and the entire vertebral column protects the spinal cord within it. The bony thorax (ribcage) forms a protective framework around the heart and lungs but also moves to allow us to breathe via 12 pairs of ribs, 12 thoracic vertebrae, and the sternum (breastbone) at the front. The sturdy pelvis protects both abdominal organs and reproductive organs and forms a large surface area for attachment of the muscles of the buttocks and lower limbs, all of them essential for locomotion (movement). Our upper and lower limbs have a similar number of bones, with a similar arrangement.

Living bones are white, their texture like ivory—tough and dense (compact, cortical bone)—or honeycombed (trabecular, spongy, or cancellous bone). Bone tissue is made up of specialized cells (osteoblasts, osteoclasts, and osteocytes) embedded within a mineralized matrix; this gives bone its strength and hardness. Embedded within the bone is a rich matrix of blood supply, which allows it to heal quickly. Around 21 million osteons (concentric cylinders of bone tissue) exist in the body; they give bone its honeycombed appearance under the microscope. Long bones are bones with expanded ends found in the upper and lower limbs. Each bone has three

parts: the expanded ends are known as the epiphysis, the narrowed neck
the metaphysis and the shaft connecting the two ends the diaphysis. During
childhood, the ends of long bones are covered in cartilaginous growth plates.
These allow growth to take place quickly. Adults do not have growth plates
on their bones.

Muscles move us voluntarily or involuntarily

Approximately 650 muscles, often arranged into layered compartments, move
our body and generate movement within every region from limbs to eyes and
even ears. Muscles attach onto the skeletal framework via tendons (or flattened
tendons known as "aponeuroses") and fascia (a type of connective tissue). Tendons
are pale and usually quite slender but strong; their blood supply is poor. Muscles,
however, are richly supplied with blood and appear red.

Muscles move air in and out of our lungs and propel food through our digestive
tract. They help us hear and see and speak and swallow—and move. Movement
is generated when a muscle contracts (or shortens), and the force a muscle
generates is entirely dependent on its shape and size. In general, short and plump
muscles contract less but may generate a lot of force; slender and long muscles
may contract a lot but generate low force. Three types of muscle tissue generate
movement within our body. *Cardiac* muscle is specific to our heart and to the
inner lining of the large vessels that enter the heart. It is not very powerful but
neither does it tire. We cannot control it consciously. Smooth muscle is within
organs, and on the inside of arteries; it is not under our voluntary control.
Muscles that move our skeletal framework are skeletal (or striated) muscles;
these muscles (such as our biceps and triceps) are quite often under our voluntary
control, but there are several skeletal muscles in our body that are normally
completely unconsciously driven. We blink, swallow, and breathe with skeletal
muscle (completely involuntarily most of the time), and the two tiny muscles
within our ears as well as those in our groin region are all skeletal muscles.

The bulk of the muscle in our body is skeletal muscle. It is capable of very
powerful contractions due to the organization of its cells (myocytes) on a
microscopic level. Skeletal muscle is under the control of the peripheral
nervous system and supplied by somatic nerves (voluntary). Both cardiac and
smooth muscles are not controlled by our consciously willing them to work. Their
activity is controlled by the autonomic nervous system, which reacts very quickly
to environmental changes and modifies our breathing, heart rate, blood flow, and
blood pressure to cope with varying demands.

LYMPH & NERVES

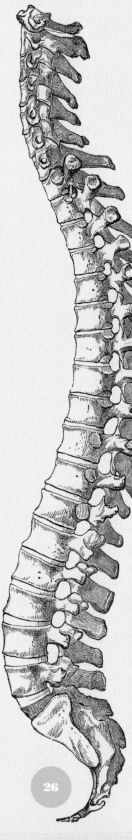

Lymph defends us

A network of vessels collect tissue fluid from spaces between the millions of cells in our body, running parallel to the circulatory system transporting blood around the body. This is the lymphatic system. Most of the lymph in our bodies is clear and colorless but that from the gut region is milky white. It forms from blood plasma (specifically, interstitial fluid) in our microcirculation and is taken up passively into lymphatic vessels from the surrounding areas. Lymph is moved along by contractile movement of the vessel walls (the larger lymphatic vessels have smooth muscle vessel walls) and by movement of surrounding muscles and arteries. Valves in the vessel walls prevent the lymph from flowing backward. The fluid needs to get back into the venous system to be recycled into the circulation and, to this end, the lymphatic vessels drain into where two large veins meet at the root of the neck (internal jugular and subclavian veins). Before lymph is deposited into the venous circulation, it goes through numerous lymph nodes dotted around the body. These small kidney-shaped nodules (between 0.1 and 2.5 cm) serve as checkpoints around the body to make sure no dangerous invaders make their way into the circulation and cause havoc. Despite their small size, they have a complex structure, are permeated with channels coursing through them and covered in a protective capsule.

The spleen and thymus gland (like the tonsils) are made up of lymphoid tissue. The cells that make up lymph tissue are known as lymphocytes. These immune cells, the gatekeepers of our immune system and part of our body's defense system, are not only contained within lymph nodes. They are scattered around the walls of the entire gut tube and also line the bronchi in our respiratory tract. Most of our immune cells (or precursors to them) are produced in the bone marrow within bones. Several lymphocytes and other immune cells as well as antibodies, the smaller immune molecules so important to our defense system, are formed within bone marrow. In a healthy young adult, there may be up to 450 lymph nodes scattered around the body. The majority of them lie close to organs in the abdomen but there are also large numbers in the groin and armpit regions, as well as in the head and neck area. Approximately 250 lymph nodes sit in the pelvis and abdomen, around 100 in the chest, and 60 to 70 in the head and neck region, amongst them a ring of tissue formed by four symmetrical pairs of tonsillar tissue that act as a first line of defense against pathogens entering our lower airways and gut.

Nerves transmit messages

The electrical superhighway of our body is the nervous system. It sends and receives messages throughout the body to control and coordinate body functions at lightning speed, achieved by perhaps as many as 100 billion interconnecting neurons (nerve cells). These neural pathways are interconnected through multiple complex networks, and information (or impulses) transmitted via electrical or chemical signaling. Voluntary and involuntary responses to information (stimuli) are triggered by the spinal cord through the network of nerves feeding into it. Every nerve cell in our body either connects to the spinal cord (neck and below) or has a pathway into the brain from the head and neck region via 12 pairs of cranial nerves. On a cellular level, each nerve has an antenna-like projection (dendrite) that receives information in a cell body, and a long extension (axon) for transmitting the information further to another cell. Nerve cells communicate with each other through chemicals they transmit within the spaces between them, triggering the next nerve cell into action.

Composed of the central nervous system (brain and spinal cord), the peripheral nervous system (cranial nerves in the head and neck and spinal nerves elsewhere—and their branches) and the sense organs (eyes, ears, nose, tongue, skin), the nervous system receives impulses through sensory nerve fibers. It interprets these impulses within the area dedicated to receiving messages arriving from a certain part of the body and then delivers a response via motor impulses to different parts of the body, either glands or muscles. The central nervous system carries out our most important functions, receiving input from sense organs and deciding how to act on the information. The peripheral nervous system is functionally two separate systems: the autonomic nervous system (involuntary) and the somatic nervous system (voluntary). The autonomic nervous system regulates internal processes in the body that keep us alive and make sure we are perfectly balanced. These processes are, for instance, our heart rate, breathing, hunger, thirst, and sweating. The somatic nervous system allows us to control our movement and even our breathing to some extent. Both divisions of the peripheral nervous system work together to maintain homeostasis (equilibrium) in the body. Within the brain, the hypothalamus monitors and regulates the state of the entire body and corrects any imbalances. The nervous system allows us to learn from our experiences and store memories.

"No knowledge can be more satisfactory to a man than that of his own frame, its parts, their functions and actions."

THOMAS JEFFERSON
LETTER TO DR THOMAS COOPER, 1814

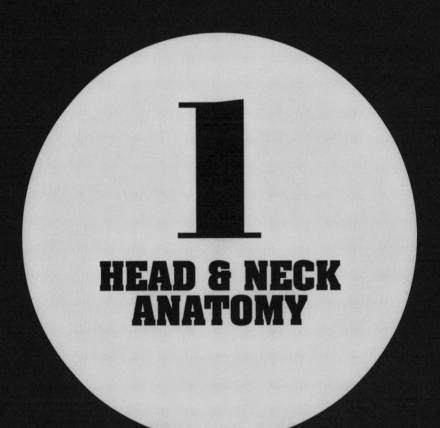

1
HEAD & NECK ANATOMY

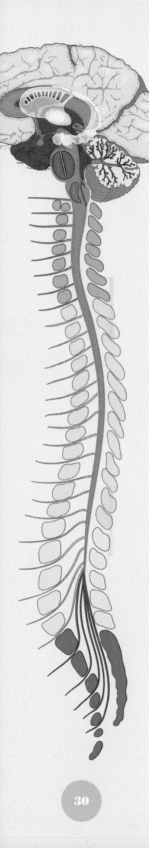

NEUROENDOCRINE FUNCTION

Our brain regulates hormonal activity between the nervous system and the endocrine system, and our internal environment is regulated and controlled by a complex, yet intricate, interplay between nerves and hormones. While, within the nervous system, nerve impulses and neurotransmitters from the autonomic system relay information very quickly to the site from which a response is required, responses via the endocrine system are slower as they travel through the bloodstream in the form of hormones, taking longer to reach their target organ or site. Hormonal responses are generally slower and more widespread, though hormones can also have an effect very quickly, if needed. Indeed, some hormones produced by the endocrine system control our emotions (say, fear or stress) and are quickly dispatched to their end destination, either a gland or tissues. The first wave of hormones triggers receptors in the target tissue to release additional hormones. A complex cascade of chemical reactions then triggers a response from the nervous system and a mechanism for expressing the emotions. This is the normal way the endocrine system interacts with the nervous system. Hormones released by glands in the reproductive system have a direct effect on the development of the nervous system.

The autonomic nervous system (part of the central nervous system) and the endocrine system are both controlled by the brain—the hypothalamus in particular, which regulates hormones. The two systems act together to regulate physiological processes in the body via a process called neuroendocrine integration. This close link between the two systems is the mechanism by which the hypothalamus within the brain maintains a steady state in the body (homeostasis). It is also the way our reproductive and metabolic functions are regulated, how we consume food and water and use this energy, and how the correct composition of chemicals inside and outside our cells is maintained. It is also the way our blood pressure is controlled.

Hormone-producing glands around the body

The endocrine system is made up of a collection of glands scattered around the body which secrete chemical messengers (hormones) through the blood (but not always). The endocrine system controls these hormones that course through our arteries. Endocrine glands are generally made up of highly active cells surrounded by a rich blood supply and tissue that loosely supports them. The cells make and secrete the hormones. To trigger an effect from another site, one single hormone is not necessarily enough, and hormones may need to trigger other hormones in a cascading chain. The hormone that starts the cascade is known as a first-order hormone (releasing hormone); second-order hormones are those that have been stimulated as a result of a first-order hormone acting on them.

Perhaps our most important endocrine gland is the pituitary gland, a small pea-sized organ on the underside of the brain measuring only about 3/8 in. (1 cm) in diameter. It is sat within a saddle-shaped depression in the floor of the cranial cavity and covered by a membrane separating it and protecting it from the rest of the contents above. Its stalk protrudes through a small hole in the membrane and is attached to the hypothalamus; together they control our most basic needs. The hypothalamus controls the pituitary gland through feedback loops that can trigger the gland to start or stop hormone production. Pituitary hormones act on numerous other endocrine glands, such as the ovaries, the testes, the suprarenal (adrenal) glands, and the thyroid gland. Hormones secreted by the hypothalamus include growth hormones, oxytocin (the "love hormone"), and antidiuretic hormone (related to blood pressure). The pineal gland is sat deep within the brain and produces melatonin, a hormone that regulates our daily cycle (circadian cycle) based on periods of natural light and darkness.

Low down at the front of the neck, clinging to the sides of the trachea and overlying it by a connecting isthmus, is the butterfly-shaped thyroid gland. Its hormones regulate metabolism. Behind the thyroid gland, four pea-sized glands—the parathyroid glands—regulate calcium levels in our body. Within the abdomen, the adrenal (or suprarenal) glands located above the kidneys produce epinephrine (also known as adrenaline), preparing the body for action or the "fight or flight" response. In close proximity is the pancreas, which produces digestive enzymes but, importantly, produces hormones that regulate our sugar levels and glucose metabolism. These are insulin and glucagon. Within the scrotum in males are the testes, which produce sex hormones but also reproductive cells (sperm). The ovaries in females are at the back of the abdominal wall and produce sex hormones and reproductive cells (ova, eggs).

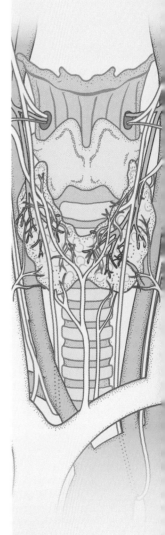

THE SKULL

GROSS ANATOMY | Mounted onto the vertebral column and upheld by the topmost vertebra (atlas), the skull protects the brain and sense organs within it and serves as an anchor for facial and masticatory muscles and ligaments. This bony skeleton of the head and face is a jigsaw puzzle of 22 individual bones (excluding six tiny ear bones), their saw-like jagged edges fusing developmentally via complex interlocking joints (sutures). The sutured bones form three small cavities (one nasal and two orbital cavities), one larger ovoid cranial cavity for the brain, and numerous smaller cavities, some of which are air-filled spaces. Only the tooth-filled lower jaw is movable; the other 21 bones are firmly knit-together. The fetal skull, though, is pliable and allows remodeling during the birthing process; incomplete bony fusion leaves soft membranous gaps that permit expansion, ensuring sufficient space for growth without damage to the rapidly growing nervous system. The skull is rarely completely symmetrical. Its sides are somewhat flattened, and its undersurface uneven and littered with numerous openings for transmitting the spinal cord, blood vessels and nerves. The smooth convex contours of the skullcap (skull vault) shielding the brain distribute any impact widely, minimizing the likelihood of fractures. The uneven facial skeleton, formed by 14 bones, exhibits wide individual variation.

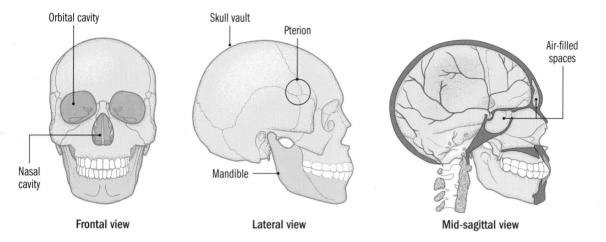

Frontal view

Orbital cavity

Nasal cavity

Lateral view

Skull vault

Pterion

Mandible

Mid-sagittal view

Air-filled spaces

CLINICAL ANATOMY | X-rays can often differentiate between the crooked line of a suture and the straighter line of a skull fracture. The midline (metopic) suture in the forehead normally disappears in childhood, giving the appearance of a uniform bone; if it remains, this suture may be mistaken for a fracture. The pterion is an H-shaped junction at the temples at which four bones interlock. Covered in life by thick muscle, the underlying bone is surprisingly thin and vulnerable; a blow can rupture the artery behind it and subsequent build-up of blood within the skull results in rapid death. Elastic young skulls yield more when injured, resulting in damage to the brain without necessarily fracturing bone.

DISSECTION | *Skull sutures are so tightly knit together they cannot be pulled apart forcefully, even after death, without damaging the multilevel interlocking system. So, anatomists used to fill the skull with dried chickpeas through its largest opening. With the skull left to soak in water over a few days, the expanding chickpeas would gently tease apart the interlocking mechanism and the individual skull bones could be separated but remain intact for later examination.*

CRANIAL CAVITY & MENINGES

GROSS ANATOMY | Two large, paired and four single bones fuse to form the cranial cavity lodging the brain. A peek into the skull reveals an uneven floor that reflects structures overlying and molding it over time. It is organized in a stepwise fashion into ever-deepening depressions (anterior, middle, and posterior cranial fossae), each housing parts of the brain, with openings (foramina) for nerves, arteries, and veins. Wedged centrally is a butterfly-shaped bone (sphenoid), its hollowed saddle-like central pit the home of the small pituitary gland, a key hormonal regulator. Bilateral rocky protrusions resembling mountain ranges (petrous temporal bone) protect the delicate inner hearing apparatus deep within it. Three concentrically arranged layers of membrane, the meninges (dura, arachnoid, and pia mater), surround the brain and spinal cord. The outermost dura mater has two layers. Its external layer loosely clothes the inside of the skull, embedding itself only into larger sutures. Its inner layers meet to form four incomplete partitions, sectioning the brain and preventing excessive movement and crushing of underlying structures but also forming spaces for venous blood to run in. Circulating cerebrospinal fluid, originating in spaces (ventricles) deep within the brain to enter a space (the subarachnoid space) between two of its layers, provides cushioning, transports nutrients, and removes waste products from the region.

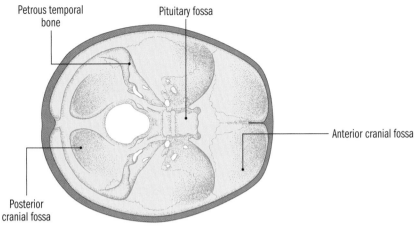

Petrous temporal bone

Pituitary fossa

Anterior cranial fossa

Posterior cranial fossa

Superior view

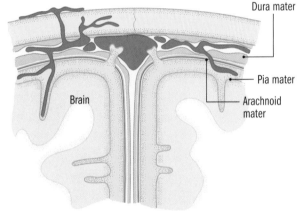

Dura mater

Pia mater

Brain

Arachnoid mater

Coronal view

CLINICAL ANATOMY | Bleeds within the rigid skull box are categorized based on their anatomical location between meningeal layers or on whether an artery within the brain ruptures or becomes blocked. The main artery supplying the dura meanders between bone and the outermost dural layer. Considerable force can damage it and result in often deadly extradural bleeds. Even mild head injuries can damage veins bridging dura and arachnoid mater, resulting in insidious subdural bleeds. Burst aneurysms bleeding into the subarachnoid space are sudden, debilitating, and often fatal subarachnoid bleeds). Blockage in or a bleed from arteries supplying the brain result in ischemic (lack of oxygen) or hemorrhagic (bleeding into the brain) strokes.

DISSECTION | *Ancient anatomists used 'mater' (mother) to indicate how structures were related in terms of physical closeness. The three meningeal layers reflect their "motherly" appearance and consistency. The leathery outer layer is the "hard mother" (dura mater). Its inner lining resembles a "cobweb-like mother" (arachnoid mater). The thin innermost lining, adherent to the brain, is the "tender mother" (pia mater).*

BRAIN & SPINAL CORD

GROSS ANATOMY | The brain and spinal cord make up the central nervous system. The brain is helpfully divided into three regions based on how the fetus grows: forebrain (cerebrum, basal ganglia, thalamus), midbrain (connecting the two), and hindbrain (cerebellum and brainstem), each connecting with the previous part. The paired cerebral hemispheres involved in higher intellectual function are the largest part, their wrinkled exterior a thin layer of gray matter, home to 100 billion nerve cells, and packing in more processing power by folding on itself. Their inner white-matter core is predominantly nerve fibers connecting areas on the same and opposite sides (via the corpus callosum) of the brain. The thalamus acts as a switchboard, relaying all sensory signals (except smell) from the body to the brain. Four paired basal ganglia are important in movement. The cerebellum regulates and maintains body movements. The brainstem is a vital center for breathing and cardiovascular function and the origin of 10 of the 12 nerves exiting the skull. While the brain resides within the closed cranial cavity, the spinal cord (around 18 in./45 cm long in an adult) journeys within the vertebral column to terminate in a bunch of nerves resembling a horse's tail (cauda equina) two-thirds of the way down the back. Through 31 sets of spinal nerves, the spinal cord transmits signals between the brain and the limbs and trunk.

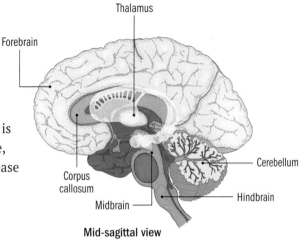

Mid-sagittal view

Forebrain
Thalamus
Corpus callosum
Midbrain
Cerebellum
Hindbrain

CLINICAL ANATOMY | The brain is well protected within a confined space, which can become problematic in disease or injury. A fast, jerking movement forward or backward (such as in a car crash) can bruise brain tissue, causing swelling and brain damage. In milder injuries, quick recovery follows a period of feeling dazed. Severe injury can result in memory loss or unconsciousness. When pressure within the rigid skull box rises following a tumor or large bleed, its contents can be pushed downward and forced through the largest opening on the floor (known as herniation or coning); damage to the respiratory center and nerves on the brainstem is inevitable and irreversible.

DISSECTION | *The unimpressive lump of soft gray tissue that constitutes the brain is nevertheless our most complex organ. Weighing a measly 3.3 lb (1.5 kg) in the average adult (2 percent of body weight) and lightened to 1¾ oz (50 g) by circulating cerebrospinal fluid, the brain is a huge energy hog. It uses 25 percent of the glucose and 20 percent of the oxygen in the blood and is insensitive to pain.*

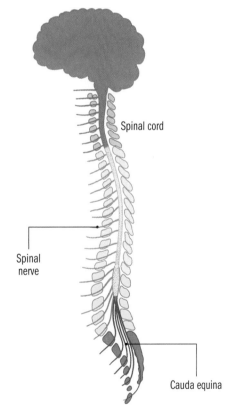

Spinal cord
Spinal nerve
Cauda equina

EYE & ORBIT

GROSS ANATOMY | The eye has one function: to transmit light to the retina to be interpreted in the brain. Seven bones form the bony orbits protecting the eyes. These resemble two four-sided pyramids lying on their sides, with their tips (apex) pointing backward and slightly inward. Their front is covered by a sheath onto which eyelids and facial muscles attach. Apart from the eyeballs, crammed into the orbits are muscles that move the eyeball, as well as vessels, nerves, lacrimal glands (for tears), and fat. Binocular vision requires synchronized movement of six eye muscles (superior rectus, inferior rectus, medial rectus, lateral rectus, inferior oblique and superior oblique), most originating at the back of the eye. The eye is lubricated externally by a continuous flow of tears that sweep across like windscreen wipers from the outer edge inward; these drain into the nose. Arranged anatomically very much like a camera, three layers encompass the eyeball: a partly white outer corneoscleral layer, a delicate, blood-vessel-laden choroidal layer and an inner light-sensitive retinal layer. At the front is a clear dome-shaped outer covering (cornea), with a lens further back to adjust focus. The iris, an adaptable diaphragm, controls the size of the opening (pupil) through which light passes to the retina. Light-sensitive photoreceptors transmit signals along the optic nerve to the back of the brain, where images are finally interpreted.

CLINICAL ANATOMY | Pressure within the eye is maintained by constant production and drainage of aqueous humor (a type of plasma). Blockage anywhere along the intricate drainage system raises pressure in the eye and can cause irreversible damage (glaucoma). As the space around the optic nerve is continuous with the space in which cerebrospinal fluid circulates around the brain, raised pressure within the cranial cavity manifests within the eye. Seen through a fundoscope, the optic nerve (optic disc) at the back of the eye appears to bulge forward, a sign not to attempt any unnecessary procedures, which may result in irreversible damage to the brain.

DISSECTION | *The transparent biconvex lens of the eye for adjusting focus has three layers to it, predominantly stretched lens cells. It has no blood or nerve supply, so that it remains clear and able to transmit light unhindered. Its surface creates a highly effective barrier against the outer world. Uniquely, it keeps all its cells for the entirety of its life.*

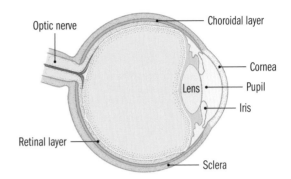

Superior view right eye

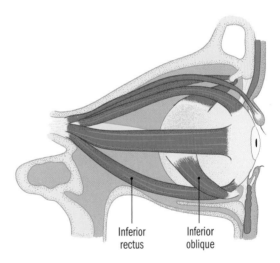

Lateral view of right orbital cavity

NOSE & PARANASAL SINUSES

GROSS ANATOMY | The nose is the irregular pear-shaped cavity in the middle of the facial skeleton. Wider at the bottom than at the top, it is bridged between the orbits by two small nasal bones. The outer nose is mainly cartilage and skin, with rich sensation and blood supply. Two front and back openings communicate with the throat (pharynx). The midline of the nose is a featureless partition, bony at the back and cartilage at the front. The sides of the nose have three bilateral shelf-like projections (each known as concha, or turbinate when covered with mucosa). Air flowing through the nostrils is cleaned via minuscule hair (cilia) and warmed via the richly supplied mucosa to flow through the throat and enter the lower airways. The soft palate below closes during swallowing to prevent backward flow from mouth to nose. Nerves at the top of the nose transmit smell into the brain through a sieve-like bone (cribriform plate of the ethmoid bone). The nasal cavity is also connected to several air-filled spaces in the adjacent bones, called paranasal sinuses, which add resonance to the voice. The largest is the maxillary sinus, above the upper teeth and below the eyes. A duct (nasolacrimal) connects the inner corner of the eyes and the area below the lowest turbinate (the inferior meatus). Tears drain here, overflowing with excessive crying.

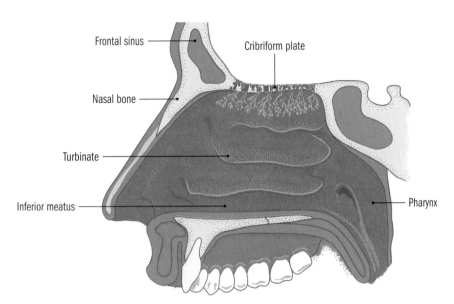

Frontal sinus

Cribriform plate

Nasal bone

Turbinate

Inferior meatus

Pharynx

Sagittal view lateral wall of the nose

CLINICAL ANATOMY | Any anatomical connections are also potential routes of infection; in the nose there are several. The paranasal sinuses often fill with infected mucus, resulting in painful sinusitis. Infection can spread from eyes to nose, and vice versa, via the nasolacrimal duct. Picking at spots in the danger area around the nose (imagine a triangle overlying the nose) can have disastrous results, as infection travels backward via veins that connect nose and eyes to the cranial cavity. The nose has a meeting point of multiple vessels on the cartilaginous part of the midline partition (Little's area), a common site of nosebleeds, which are often the result of nose-picking.

DISSECTION | *Gravity and breakdown of cartilage affect our aging bodies, not least the nose and ears, which appear to continue growing into old age. This is an illusion. Due to age-related changes in the cartilaginous framework (the tip starts to weaken and sag) and facial bones (the maxilla shrinks), the shape of the nose changes. Viewed in profile, it appears lengthened.*

SCALP & FACE

GROSS ANATOMY | Anatomically, the face does not include the ears. It extends from forehead to chin, between the ears. The facial skeleton, including the movable mandible (jaw), has 14 bones (6 paired and 2 single). Two paired orbital cavities, a pear-shaped nasal cavity, and the oral cavity house the sense organs of sight, smell, and taste. Two sets of muscles sit within the face: masticatory muscles (chewing) and facial muscles. The approximately 19 muscles forming the latter group have a dual purpose. They are arranged around the openings of the facial skeleton into dilators and sphincters for opening and closing the eyes, nose and mouth, their names often reflecting their action. Known as mimetic muscles, they enable a range of human emotion to be expressed. The scalp is continuous with the face; anatomically, it starts from above the eyebrow ridge in the forehead and extends to the back of the skull; sideways, it reaches down to the cheekbone and the bony opening of the ear canal. Composed of five layers, its top three layers move as a unit. The third layer of this unit is a double-bellied muscle layer (also a facial muscle) separated by a flattened tendon (aponeurosis). The innermost layer is tightly bound to the skull.

CLINICAL ANATOMY | Due to its richly overlapping blood supply, even small scalp injuries bleed profusely and appear dramatic. As the majority of the blood and nerve supply is within the fatty layer just below the surface (and the three topmost layers move as one), sections of skin including blood and nerve supply (flaps) can be grafted elsewhere for corrective surgery. Deeper cuts to the scalp gape and bleed horrifically as the two muscle bellies separated by the aponeurosis pull in opposing directions; careful suturing and re-approximation of the layers stem the flow. A small arterial connection between the temple and the eye can sometimes spread inflammation of the arteries in the face region (temporal arteritis) to the eye, causing sudden-onset blindness.

DISSECTION | *Unusually, facial muscles insert into each other and the overlying skin, allowing a wide range of movement. Platysma, originating from the chest and sweeping upward over the neck to join muscles in the face, is a rare example in humans of a thin muscular sheet, (panniculus carnosus) connected to overlying skin, which in other mammals is used to flick off a pesky insect.*

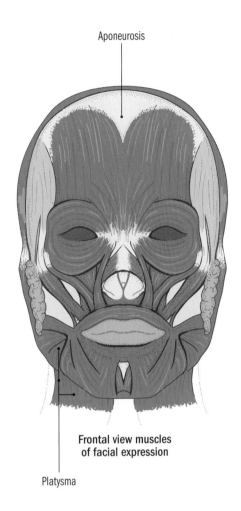

Aponeurosis

Frontal view muscles of facial expression

Platysma

MOUTH & TONGUE

GROSS ANATOMY | The oral cavity starts at the lips and cheeks and extends inward via a narrow oropharyngeal isthmus into the throat (pharynx). Its primary function is for food consumption, but it is also a passageway for ventilation and enables speech. The area in front of the teeth bounded by the lips and cheeks is the vestibule. Behind the teeth, the palate separates the mouth from the nose; the arched soft palate, with its midline fleshy uvula, closes the passageway in the throat between mouth and nose and facilitates swallowing and speech. Salivary glands lubricate the mouth. A hammock-like muscular arrangement on the floor snugly accommodates the tongue which, covered in moist pink mucosa and attached to several bones above and below it, is partly in the mouth (apex and body), partly in the throat (root). It is a fusion of eight paired muscles, separated in the midline by a fibrous partition. Four extrinsic muscles attach to different areas outside the tongue to enable large movements. Four intrinsic muscles, entirely within the tongue, alter its shape. Most of the tongue's surface is covered in three types of small taste buds, especially plentiful at the back. Its undersurface is tethered below by a fold of tissue (frenulum). Five of the 12 cranial nerves supply the tongue, illustrating its importance.

CLINICAL ANATOMY | The double-arched oropharyngeal isthmus acts as a passageway for ingested substances but also as a gatekeeper for preventing harmful substances entering the lower airways. Tonsillar tissue lying in a mucosal bed between the arches bulges into the isthmus from either side. Most infections in the palatine tonsils are self-limiting but, occasionally, when the tonsils meet in the midline (kissing tonsils), there is little room for food and water to travel to the pharynx. An episode of tonsillitis may present as earache (or vice versa), as the middle ear and the palatine tonsils share a nerve (glossopharyngeal). Pain in the throat or ears requires examination of both areas.

DISSECTION | *The set of upper and lower teeth within the mouth are virtually indestructible and are used for forensic identification when the face has been destroyed. They are also far more reliable for determining true age than bones, even in the developing fetus. At the root of the teeth, the amount of translucent dentine deposited is more or less proportional to age.*

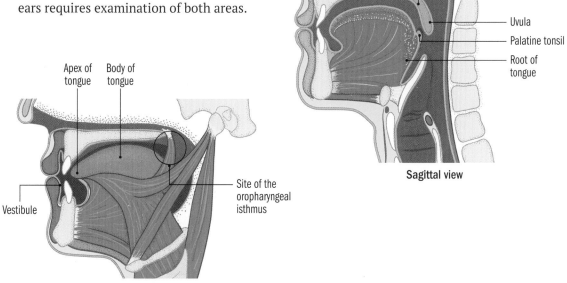

Hard palate   Soft palate

Uvula
Palatine tonsil
Root of tongue

Sagittal view

Apex of tongue   Body of tongue

Vestibule

Site of the oropharyngeal isthmus

Sagittal view

SALIVARY GLANDS

GROSS ANATOMY | Three paired major salivary glands (parotid, submandibular, and sublingual), and 800 to 1,000 minor ones scattered around the oral cavity, produce saliva to moisten and lubricate food, initiate digestion, prevent tooth decay, and enable speech through moistened mucosal surfaces. The largest is the parotid gland in front of the ears. Covered by a fibrous capsule, three structures run through it (artery, vein, and nerve). The facial nerve passes forward through the gland to supply the muscles of facial expression and divides it into two lobes. The submandibular gland is wedged under the angle of the lower jaw. Its two lobes, like a horseshoe, tuck around the edges of a hammock-like floor of the mouth muscle (mylohyoid). Closer to the midline, underneath the tongue's mucous membrane, are the almond-shaped sublingual glands. The parotid gland produces a serous (watery) secretion, the sublingual gland a mucous secretion, and the submandibular gland a mixture of these. All major glands empty saliva into the mouth via ducts, the openings of which are visible to the (trained) naked eye. The parotid empties its contents through a duct that opens in the upper cheek just opposite the second-last tooth at the back (upper second molar) via the parotid duct; saliva from the submandibular, and sublingual glands empty out via openings underneath the tongue via Wharton's duct.

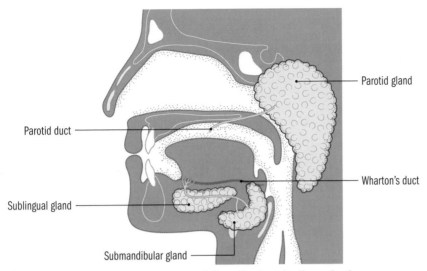

Parotid gland

Parotid duct

Wharton's duct

Sublingual gland

Submandibular gland

Schematic view of salivary glands

CLINICAL ANATOMY | The capsule surrounding the parotid gland can barely stretch and the gland is limited in three directions. Conditions affecting the gland (such as mumps) cause severe pain as the capsule is stretched. Parotid swellings tend to extend into the neck as the only potential direction of growth. Parotid tumors compress the facial nerve running inside the gland, causing drooping (palsy) of the muscles of facial expression on the affected side. Bell's palsy indicates an unknown cause for this phenomenon. If the forehead muscles are still functional, the cause is likely to be within the brain, as the forehead facial muscles receive two sources of nerve supply and are spared when problems occur with the brain.

DISSECTION | *Every day, around 3 pt (1.5 l) of saliva is produced at a steady rate of 0.01 fl oz (0.3 ml) per minute. The salivary glands are stimulated merely at the thought, sight, or smell of food, and flow can rise to 0.05 to 0.07 fl oz (1.5 to 2 ml) per minute, with the parotid gland contributing to about 50 percent of saliva produced. Barely any saliva is produced during sleep.*

MANDIBLE & MASTICATION

GROSS ANATOMY | Buttressed in the midline to form the chin, the U-shaped mandible has a horizontal arch in which vessels, nerves, and teeth are contained. The only movable joints in the skull are on the sides of the face where the vertical projections of the mandible meet the temporal bone of the skull (temporomandibular joint). The mandible is the attachment site for the muscles of mastication (chewing). Four paired muscles enable forward sliding (protrusion), side-to-side, lifting up (elevation) and pulling back (retraction) movements. To open the jaw, a muscle (lateral pterygoid) pulls the head of the mandible forward so the jaw is slightly protruded. After this, gravity takes over to open the jaw (when upside down, the neck muscles can open the jaw). From the open position, the other three muscles lift up and pull back the mandible to enable speech and chewing. Two of the muscles can be felt if the jaw is clenched, one overlying the temple (temporalis) and the other overlying the vertical projections of the mandible on the lower side of the face (masseter). The deeper muscles sit in a crowded deep space called the infratemporal fossa, which contains nerves and arteries and a meshwork of veins with numerous connections within the face and into the skull.

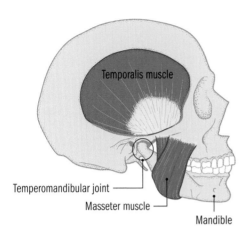

Temporalis muscle

Temperomandibular joint

Masseter muscle

Mandible

Lateral view

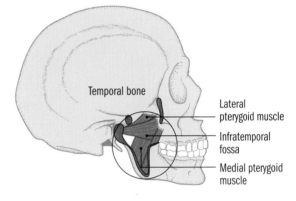

Temporal bone

Lateral pterygoid muscle

Infratemporal fossa

Medial pterygoid muscle

Sagittal view with mandible partly removed

CLINICAL ANATOMY | The temporomandibular joint at the side of the skull is a modified suture, with the features of other joints (capsule, cartilage). The lateral pterygoid muscle initiating jaw opening (or the nerve supplying it) can become affected in deeper infections (usually collections of pus) around the mucosa surrounding the palatine tonsil (quinsy) or within the infratemporal fossa, in which it lies, or when dental anesthesia injected into the oral cavity has slightly overshot. In these instances, spasm (trismus) follows and opening the jaw becomes virtually impossible. Sometimes, the head of the mandible within the joint cavity can move too far forward and become dislocated, with mouth agape.

DISSECTION | *The chin is unique to humans. At birth the mandible is in two parts, united in the midline by a fibrous suture. Fusion occurs over the first three years. Genetic inheritance determines formation of a cleft chin (fovea mentalis), indicating incomplete fusion of the midline suture. These dimples can also occur if the overlying mentalis muscles (facial muscles) fail to unite.*

THE EAR

GROSS ANATOMY | Easily visible in varying shapes and sizes on either side of the head, the majority of the ear is, however, embedded within the temporal bone below the floor of the cranial cavity. It is divided into external, middle, and inner parts, all crucial in the complex process of hearing and balance maintenance. The external ear consists of the grooved cartilaginous auricle, which collects sound waves, and the part-cartilaginous, part-bony external auditory canal, which guides sound to the middle ear. When sound passes through the ear canal, it vibrates the eardrum, a translucent oval membrane partitioning the external from the slit-like middle ear. The vibration moves a chain of three tiny bones (ossicles) conducting sound to the labyrinthine fluid-filled cavities and passages of the inner ear. Two functional units coexist within the inner ear: the vestibular apparatus maintains balance, some parts of it detecting gravitational forces and positions of the head; the snail-like cochlea is for hearing. The last in the series of tiny bones in the middle ear knocks on a membranous window of the cochlea, causing the fluid within to move, triggering a response in the nerve supplying the area. Sounds are interpreted in multiple locations within the brain.

CLINICAL ANATOMY | The ear is vulnerable to damage due to its structural complexity, cramped quarters, and relationship to adjacent regions. Auricular skin is easily separated from underlying cartilage to which it is tethered. Prompt treatment to external ear injuries aims to reconnect cartilage with overlying nutrients and prevent "cauliflower" ear. Throat infections reach the ear via a ventilation passage that equalizes air pressure on either side of the eardrum; in children, the resulting middle ear infection can spread to air-filled bone (mastoiditis), with severe consequences. Damage to stapedius, the smallest muscle in the body (less than 1/16 in/1 mm) that dampens sudden and high-frequency sounds, results in normal sounds being perceived as extremely loud.

DISSECTION | *The smallest bones in humans are the three middle-ear ossicles, shaped as hammer (malleus), anvil (incus), and stirrup (stapes), two of which have minuscule muscles attached to them. Uncommonly for bone, ossicles are thought to be more or less fully developed at birth, providing babies with similar sound transmission to adults and enabling normal speech and language development.*

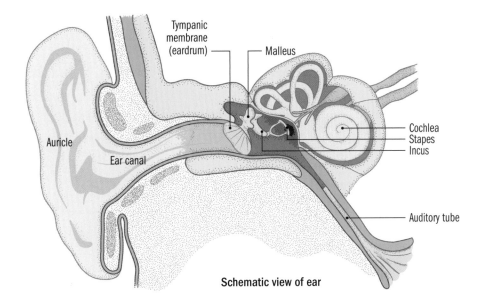

Schematic view of ear

NECK TRIANGLES & COMPARTMENTS

GROSS ANATOMY | The neck extends from the bottom of the mandible to the top of the sternum (breastbone), connecting the head with the rest of the body. It supports the head via the vertebral column at the back, its shape taking form around this and the trachea in front. It is relatively long in humans and highly flexible; rotation and side-to-side movement of the head happen at the upper two vertebrae (atlas and axis). Organized into four longitudinal fascial compartments, the structures within are protected and can move in opposing directions unobstructed. The largest compartment is surrounded by a thick double-layered investing fascia that wraps around the neck like a collar and splits around muscles it encounters. The other three compartments are within it, wrapping their valuable cargo in further fascial layers, not unlike packing material. The sternocleidomastoid muscle lying obliquely on either side of the neck divides the neck into triangular spaces (anterior and posterior triangles). The U-shaped hyoid bone high in the neck further divides the front triangles into suprahyoid and infrahyoid regions. Four bilateral suprahyoid muscles lift the bone upward or sideways to aid in opening the mouth; four bilateral infrahyoid muscles tend to pull down the hyoid as well as the larynx to which it is attached.

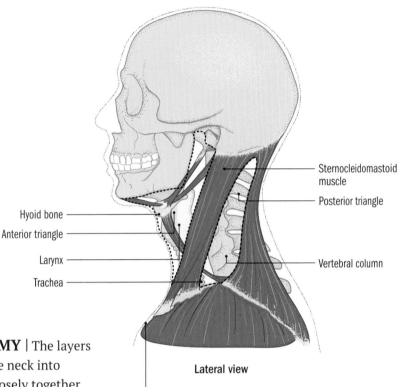

Hyoid bone

Anterior triangle

Larynx

Trachea

Sternocleidomastoid muscle

Posterior triangle

Vertebral column

Lateral view

Sternum

CLINICAL ANATOMY | The layers of fascia separating the neck into compartments cling closely together with no physical space between them. They have the potential to expand and spread infection around the larynx and pharynx, into the chest, and even as low down as the back. An infection from an impacted tooth can spread beyond the floor of the mouth. Pus filling the space pushes structures, such as the larynx and trachea, and can obstruct breathing and lead to death. Neck lumps occur within predictable locations within the neck triangles. A small midline outpouching could be a cyst from a duct that has failed to close during thyroid development.

DISSECTION | *Humans do not have the same anatomical flap of skin (or dewlap) dangling from the neck as some other species (cows and turkeys). The collar-like investing fascia of the neck, attaching firmly from hyoid bone to mandible, prevents anatomical dewlap but not the pendulous loose skin (or double chin) that makes its unfortunate appearance as we age or expand.*

LARYNX & PHARYNX

GROSS ANATOMY | The larynx is the gateway to the lower respiratory tract. Acting like a sphincter (a muscle that constricts) to prevent undesirable substances from entering, it only allows air into the trachea. It lies below the hyoid bone in the midline of the neck, and immediately behind skin, fat, and a thin sheet of muscle (platysma). Its framework (several cartilages, ligaments, and membranes) is attached to the hyoid bone; they move together. Mucosa covers everything. Underlying muscles widen or narrow the gap between the vocal cords to allow air to pass through or to produce sound. Only one muscle opens this gap. By closing the gap completely, the larynx can stabilize the chest just before straining or lifting. Covered in five layers of mucosa, fascia, and muscle, the 5 to $5\frac{1}{2}$ in. (12 to 14 cm) long pharynx hangs from the base of the skull behind the nose, mouth, and larynx (naso-, oro- and laryngopharynx). It resembles a tube, with three openings in the front for communication with these spaces. It directs air via the larynx into the respiratory tract and channels food from the mouth to the digestive tract via three circular muscles that propel food down. Three longitudinal muscles lift and open it up. The pharynx also helps with vocalizing sounds and equalizing pressure within the middle ear.

CLINICAL ANATOMY | A branch of a nerve supplying the larynx travels through the neck to the chest, makes a U-turn, and returns to supply the vocal cords. A lung tumor pressing on it can result in voices changes; any voice changes warrant an examination of the lungs. Tonsillar tissue in the throat (back of the nose, mouth, and tongue) forms a protective ring, a first-line of defense for the lower respiratory tract. Foreign objects often get lodged in spaces within the laryngopharynx, the pear-shaped piriform recess being a common site. Tumors can also grow to a large size undetected in this space before pressing on vital structures.

DISSECTION | *The laryngeal ventricle is a narrow space between false (upper) and true (lower) vocal cords. This space extends into saccules supplying lubricating material for the cords. Other primates, notably howler monkeys, have inflatable saccules, but in humans they are insignificant. Wind instrument players and those prone to heavy physical work may have enlarged saccules, fortunately imperceptible to the eye.*

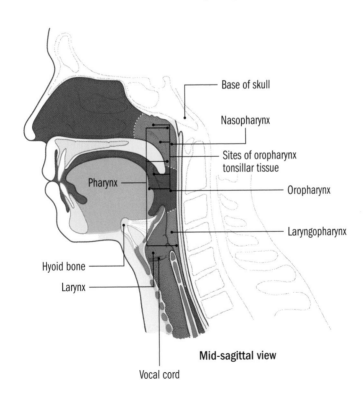

Base of skull

Nasopharynx

Sites of oropharynx tonsillar tissue

Pharynx

Oropharynx

Laryngopharynx

Hyoid bone

Larynx

Mid-sagittal view

Vocal cord

THYROID

GROSS ANATOMY | The thyroid and parathyroid glands are endocrine glands found low in the front of the neck, overlying the trachea and below the larynx. The thyroid gland is relatively large and unpaired. It secretes hormones essential for growth and metabolism and stores its secretions outside the mass of cells it is made from, something unique among all the endocrine glands. It has two lobes, snugly hugging the trachea, which are connected by a horizontal narrowing (isthmus). The entire gland is covered by a capsule, which clings to the trachea behind it by a small fascial attachment. On swallowing, the trachea rises and the thyroid gland moves along with it. Four pea-sized parathyroid glands, each one weighing around 1¾ oz (50 mg) and secreting parathyroid hormone essential to calcium regulation, can normally be located behind the thyroid gland, although they vary in number and location. They are enclosed within the capsule surrounding the thyroid gland or behind it. Both glands lie within an additional outer fascial layer (pretracheal fascia) covering them and the trachea. So vital are these two glands that their blood supply comes off two major arteries (carotid and subclavian). In rare individuals (3 percent), the thyroid gland receives an additional artery (thyroid ima) directly off the aorta.

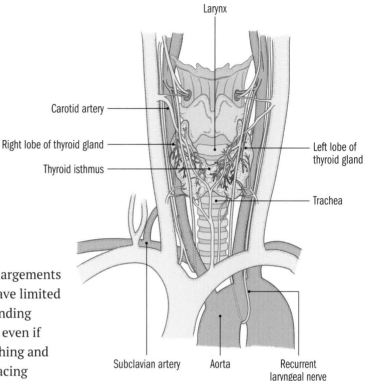

Larynx

Carotid artery

Right lobe of thyroid gland

Thyroid isthmus

Left lobe of
thyroid gland

Trachea

Subclavian artery — Aorta — Recurrent
laryngeal nerve

Anterior view thyroid gland

CLINICAL ANATOMY | Enlargements
of the thyroid gland (goiters) have limited
growth space due to the surrounding
capsule. Some require removal, even if
benign, as they can cause breathing and
swallowing difficulties by displacing
the trachea and esophagus. Cancerous
growths do not displace; instead, they
erode these structures. Close to the
gland are nerves that supply the larynx
(recurrent laryngeal nerve) and a chain
of nerves that starts in the brain but
makes a loop in the neck to travel to the
face (sympathetic chain). If compressed
or eroded, voice changes or a subtle
drooping of an eyelid may be noted.
Thyroid surgery requires finding the
thyroid ima artery and protecting it,
or a major bleed can follow.

DISSECTION | *Curiously, the thyroid
gland starts its developmental journey at
the floor of the pharynx and pushes its way
through the tongue to travel to its rightful
place at the front of the neck. Occasionally,
it overshoots and migrates into the chest.
Very occasionally, a completely functional
thyroid gland overlies the tongue, having
decided never to migrate in the first place.*

"Anatomy is the foundation of medicine and should be based on the form of the human body."

HIPPOCRATES

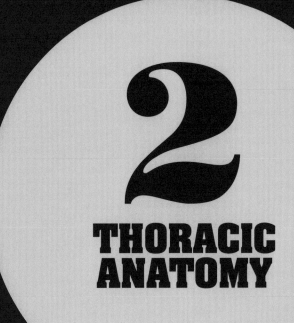

2
THORACIC ANATOMY

RESPIRATION & CIRCULATION

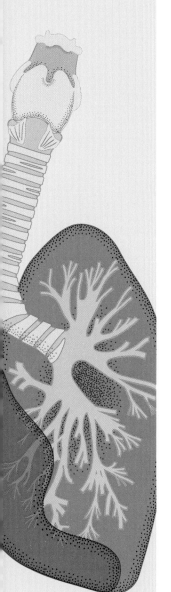

All our cells are dependent on a regular and constant supply of oxygen to survive. At each cell, oxygen is exchanged for carbon dioxide, which needs to be removed from the body as it is harmful in high doses. Although our blood transports these gases around the body, the ultimate place where gases are exchanged (gaseous exchange) is in the lungs. Air arriving in the lungs is full of oxygen. When it reaches the very small sacs in the lungs (of which there are millions), there is only a very thin membrane at that level—so thin that the gases glide across easily into minuscule blood vessels (capillaries) and from there into bigger blood vessels and into the heart to be sent to tissues. Fresh oxygen is exchanged for carbon dioxide, which is then pushed out of the lungs to be exhaled. This is respiration, or breathing. The shuttling of blood around the body is done by the cardiovascular/ circulatory system, which is heavily dependent on the respiratory system.

The body has its own one-way muscular pump, the heart. Together, the heart, blood, and blood vessels make up the cardiovascular system. The heart squeezes (contracts) to push the blood within it through an interlinked network of vessels. By doing this, it delivers life-giving oxygen and nutrients to every cell in your body. Through the bloodstream, hormones needed to regulate growth, metabolism, and other vital functions travel to their target areas. White blood cells, circulating to pick up any invaders that have got past other defense systems, also course through our vessels. Without constant blood flow to and from every cell in the body, to supply them with oxygen and nutrients as well as remove waste products from them, our bodies would not last more than a short period of time before cell death kicked in. The blood carries waste products to the liver and kidneys for excretion. Ultimately, the circulatory system is made up of the heart, blood, and blood vessels, but other organ systems (nervous and endocrine) also directly and indirectly regulate it.

Blood vessels under the microscope

Blood vessels—arteries, arterioles, capillaries, venules, and veins—vary in size throughout the body. The vessel walls of arteries and arterioles are thick and elastic and designed to help them cope with blood being pumped through them at high pressure via the systemic circulation, much like the high volume of traffic on a motorway. Arteries and veins are made of three layers (tunics), each layer of which has several sublayers. The innermost layer is the tunica intima, a thin layer of endothelial cells that lines the entire circulatory system. The middle layer is the tunica media, a muscular wall. Within arteries, the tunica media is made up of more smooth muscle than in veins, which allows them to widen (dilate) and narrow (constrict) to adjust the amount of blood needed by the tissues. This vasoconstriction (narrowing) is the mechanism by which the body regulates blood pressure and temperature. The outermost layer of the vessel walls is the tunica adventitia, which is a thick layer of connective tissue. Capillaries are one-cell-thick vessels at the intersection between the arterial and venous systems. They allow exchange of oxygen, carbon dioxide, and other nutrients and wastes between the two systems.

Pulmonary and systemic circulation

Two types of circulation are ongoing at any one time within our body. Blood low in oxygen that has been returned to the heart from the rest of the body is pumped from the right side of the heart to the lungs, where it is infused with oxygen and then returned to the left side of the heart. This is the pulmonary circulation, a low-pressure system ensuring fluid from the small vessels around the lungs (capillaries) is not forced into the little air-filled sacs in the lungs (alveoli). Blood is returned from the lungs to the left side of the heart and shunted at high pressure to the rest of the body, the systemic circulation. This blood is now rich in oxygen and thus life-giving. Pressure in the systemic circulation is high so that blood can be pushed to every area of the body to supply organs as well as the extremities of our bodies: our toes and fingers.

Every major blood vessel in the body originates from the aorta, the superhighway of blood vessels. It first gives off vessels to supply the heart (right and left coronary arteries) and three major arteries that travel upward to supply the head and neck and the arms. As the control center of our body, the brain requires priority blood supply. Second only to the heart itself, the brain is always targeted to receive blood quickly and efficiently. Oxygen-rich (arterial) blood arrives rapidly via a dual-supply consisting of the internal carotid and vertebral arteries, which form a network of communicating arteries on the underside of the brain (the "circle of Willis").

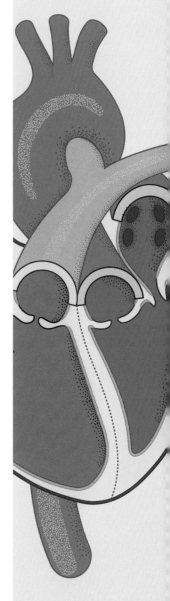

THE BONY THORAX

GROSS ANATOMY | The chest cavity is the second-largest hollow space in the body and houses three independent spaces within it for the heart and two lungs. It is separated from the abdominal cavity below by the muscular diaphragm. Its contents are enclosed by a chest wall made of skin, fat, muscles, and the bony ribcage. Stripped of all its coverings, the bony thorax resembles a cage with a narrow upper opening and a wider lower opening. This sturdy structure protects the contents within but allows expansion for breathing in two directions: sideways expansion resembles a bucket handle being lifted up and down, and upward movement the action of an old-fashioned pump handle. So that the organs within are not damaged, other movement is limited to only a small amount of rotation. The ribcage is formed by the vertebral column (spine), ribs, and sternum (breastbone). Ribs give the chest its shape. Twelve ribs on either side of the vertebral column slope downward and angle toward the dagger-shaped sternum onto which the majority of them attach. The top seven ribs (true ribs) join the sternum directly via cartilaginous endings. Ribs eight to ten (false ribs) join the sternum indirectly by attaching onto the cartilage of the rib above. Two floating ribs have no attachment onto the sternum and only attach onto the spine.

CLINICAL ANATOMY | Children rarely fracture their ribs as their ribcages are very elastic. In adults, rib fractures happen commonly despite fairly flexible ribs, the most vulnerable part being in front of where a rib curves forward. The clavicle protects the top two ribs, and the freely swinging lower two ribs are also less likely to get fractured. Individual rib fractures are usually allowed to heal on their own without intervention, although when multiple ribs are broken, they tend to get treated due to their potential for damaging the underlying lungs. The sternum rarely gets fractured as the ribs attached to it make it quite springy and difficult to break.

DISSECTION | *The ribcage takes its adult shape and proportions when a child begins to walk, before which it appears splayed out. Women have a shorter sternum and the upper opening of the ribcage slants forward at more of an angle than in men. They have smaller ribcages than men and the capacity of their lungs to expand is also less.*

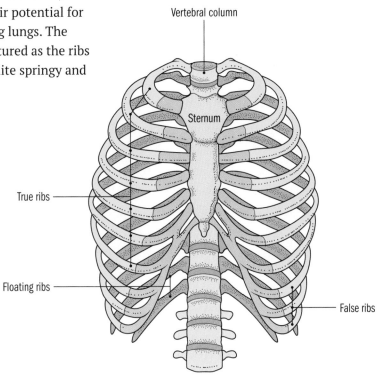

Vertebral column

Sternum

True ribs

Floating ribs

False ribs

BREAST

GROSS ANATOMY | Breasts (mammary glands) are modified sweat glands. They sit on top of the large muscles in the chest (pectoralis major) and the ribcage, and secrete milk in women who have had a baby. Their development is regulated by several hormones. Estrogens grow breast cells, progesterone differentiates these into milk-producing cells, and prolactin and somatomammotropin induce milk production. After menopause, when these hormones stop being produced by the body, the breasts shrink. In women, the breasts are formed into smooth contours by 15 to 20 lobules of glandular tissue, embedded within fat. The lobules are separated into smaller fibrous compartments, suspended via bands from the skin to the chest wall. Milk drains from the lobules into the nipple via ducts. Milk formation takes place in small honeycomb-like pockets (alveoli) sprouting from the ducts. These increase greatly during pregnancy and secrete fatty milk droplets, which collect into milk reservoirs beneath the nipple. Roughened pigmented (darkened) skin surrounds the nipple at the areola, a region containing numerous sweat and sebaceous glands. During pregnancy the areola darkens intensely, and this change in pigmentation may persist through life. In men, male hormones suppress breast development during puberty, so they have undeveloped breast tissue and no anatomical capacity for producing breastmilk.

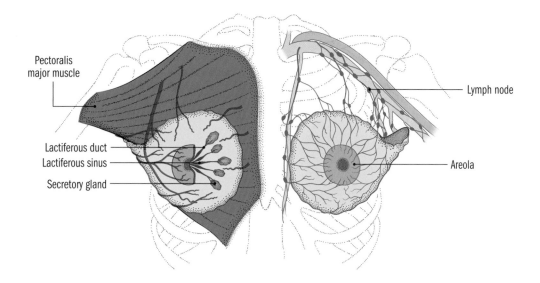

Pectoralis major muscle

Lactiferous duct

Lactiferous sinus

Secretory gland

Lymph node

Areola

CLINICAL ANATOMY | In breastfeeding women, bacterial organisms can invade the deeper breast tissues through the nipple and cause intensely painful mastitis, an infection unique to the breast. Problems with the endocrine regulation of hormones can result in a condition called precocious puberty, where a young child appears to have adult breast tissue. This can affect both boys and girls, and may be caused by structures affecting the hypothalamus or pituitary gland in the brain (central causes) or those affecting the gonads (peripheral causes). In young boys, elderly males, or those on medication that affects male hormones, a hormonal imbalance can cause one or both breasts to become enlarged (gynecomastia).

DISSECTION | *The breasts are a fairly common site for developmental anomalies. In rare cases, a breast may be missing altogether (amazia) or it may be very small on one side (the left is usually larger) or both sides. Occasionally, extra nipples (supernumerary nipples) or extra breasts may occur along vertical "milk" lines, not dissimilar to the mammary glands of other mammals.*

THE LOWER AIRWAYS

GROSS ANATOMY | The lower airway passage resembles an upside-down tree with a large trunk and numerous large and small branches. Oxygen makes its way through the upper airways into the lungs via the trachea and two main bronchi. The trachea starts in the neck below the larynx and cricoid cartilage and travels down vertically in the midline of the neck, where it can be felt from the outside. It is a highly mobile structure covered in mucous membrane, whose position and length changes during breathing. In the average person, the approximately 4 in. (10 cm) long trachea is made up of 16 to 18 C-shaped incomplete rings of cartilage, which keep it open so air can flow through unobstructed. Its back is a flat fibrous wall that bulges inward into the trachea when food passes down the esophagus immediately behind it. The trachea splits into the two main bronchi in the upper part of the chest, with one entering the left lung horizontally (the heart is below it) and one entering the right lung almost vertically. Air is channeled down ever-narrowing tubes (multiple narrower bronchi and bronchioles) into small air-filled pouches (alveoli) at which gas exchange between blood and air takes place. Oxygen is taken up by the body and carbon dioxide removed.

CLINICAL ANATOMY | The arch of the aorta leaving the heart loops over the left main bronchus. A weakened and bulging aortic vessel wall (aneurysm) is at risk of rupturing and pulls on the bronchus and trachea regularly as a result of breathing. This tracheal tug is visible in the little dip just above the breastbone and is a rare sign of underlying disease. The wider, shorter, and more vertical right main bronchus is a common entry route for objects inhaled into children's lungs. If the upper airways are obstructed, access to the lungs can be gained precisely in the midline through the trachea (tracheostomy) to avoid major vessels in the neck.

DISSECTION | *A thin film with detergent-like properties (surfactant) covers the surface of the nearly 500 million tiny thin-walled balloon-like alveoli that cover the surface area of 5050 cu ft (143 m³) in an adult's lungs. Surfactant minimizes surface tension and prevents the alveoli from collapsing, making ventilation more effective. Premature babies are given artificial surfactant to help them breathe.*

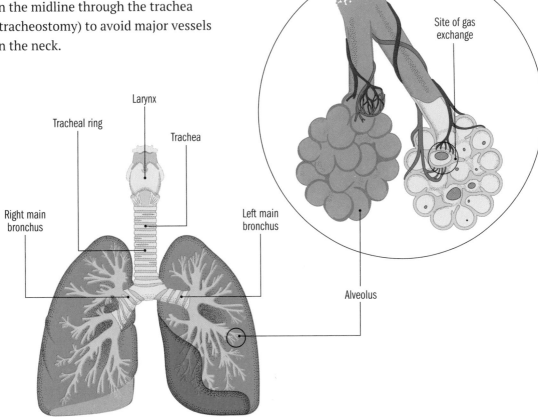

Larynx

Tracheal ring

Trachea

Right main bronchus

Left main bronchus

Site of gas exchange

Alveolus

LUNGS & PLEURA

GROSS ANATOMY | The lungs are the vital organs for breathing. They shift oxygen soaked up from the air into blood cells and exchange it for carbon dioxide, removed with every outbreath. The right lung has three lobes (superior, middle, inferior) separated by two deep slits (horizontal and oblique fissures). The left lung has two lobes (superior, inferior) separated by an oblique fissure. A narrow cone-like apex projects above the first rib and can be felt in the depression behind the clavicle. Each lung is virtually free-floating and sits within an independent cavity inside the larger chest cavity, covered by a thin membrane (visceral pleura), not unlike plastic wrap, which dips into its slits and cracks and is tightly stuck onto the surface of the lung. This layer folds back on itself and forms a second outer layer, which completely covers the inner surface of the chest wall (parietal pleura). The pleural cavity between the layers contains only a few milliliters of fluid, which oozes out to moisten and lubricate the pleura, and is under vacuum-like negative pressure. The layers slide over each other like two sheets of glass, separated by a few drops of water, which can slide in relation to each other but cannot be pulled apart.

CLINICAL ANATOMY | Conditions that affect the lung coverings or the space between them cause difficulty in breathing or painful breathing. Pleurisy is painful, widespread inflammation of the pleural membranes. A breach of the lung coverings (from rib fractures or operations) can lead to collections of air (pneumothorax), lymph (chylothorax), blood (hemothorax), or fluid (pleural effusion) between the visceral and parietal pleurae. The space between the membranes (normally in close contact) increases, pushing the deeper lung tissue aside and compromising breathing. Chest drains are placed through the chest wall into the pleural cavity to remove substances and relieve the pressure on the lungs, to allow for full expansion.

DISSECTION | *Freshly removed healthy lungs float in water, are spongy, and crackle when handled due to air within the alveoli. Badly diseased lungs sink. Adult lungs are blotchy and gray, with carbon deposits from the air blackening them even further with advancing age. Men (more than women), smokers, and those living in industrial areas have blacker lungs. Babies' lungs are pink.*

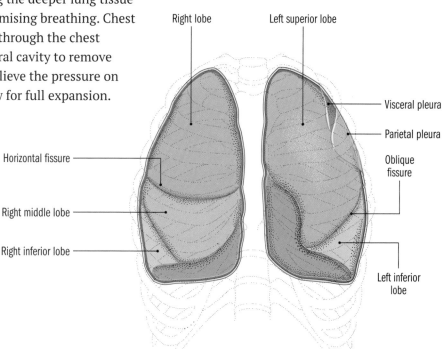

Right lobe

Left superior lobe

Visceral pleura

Parietal pleura

Horizontal fissure

Oblique fissure

Right middle lobe

Right inferior lobe

Left inferior lobe

MUSCLES OF RESPIRATION

GROSS ANATOMY | Most breathing takes place with little effort and no conscious thought. With every breath in, the ribcage gently moves upward and outward and the diaphragm flattens a little (quiet inspiration). This is followed by the elevated ribs recoiling back to their original position and the diaphragm passively relaxing with breathing out (quiet expiration). This can be overridden by voluntary control of breathing (which can be forced inspiration and expiration). The parachute-shaped diaphragm is the main muscle for breathing and does not tire easily. It separates the chest cavity from the abdominal cavity; the only communication between the two spaces is through three openings associated with the diaphragm allowing esophagus, inferior vena cava, and aorta through. The edges of the diaphragm are muscular and its center a flattened tendon. Within the chest cavity, it is firmly attached to the pericardial sac; the heart moves up and down with breathing. Between the rib spaces (intercostal spaces), three layers of muscles crisscross in opposing directions and maintain the shape of the chest wall as well as shrinking or expanding the ribs with breathing. Other muscles in the neck, back, chest, and abdomen also aid in breathing, especially during forced inspiration or expiration when a lot of oxygen is needed.

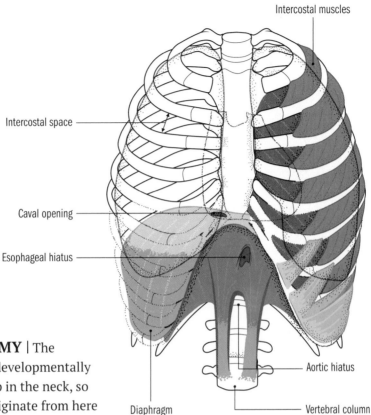

Intercostal muscles

Intercostal space

Caval opening

Esophageal hiatus

Aortic hiatus

Diaphragm

Vertebral column

CLINICAL ANATOMY | The diaphragm starts out developmentally as a small inverted cup in the neck, so its nerves (phrenic) originate from here and travel a long course through the thorax to reach their destination, piercing through the diaphragm to supply it from its underside in the abdominal cavity (not from the top, as expected). Other nerves originating in the neck supply areas over the shoulder. Any irritation to the phrenic nerve endings in the abdomen (a large bleed) or along the course of the nerve (a tumor) is felt in the shoulder, as the brain gets confused by mixed messages converging at the same levels of the spinal cord.

DISSECTION | *Hiccups are involuntary spasm-like contractions of the diaphragm. When a spasm occurs, a sudden intake of breath is cut short as the space between the vocal cords closes tightly, producing the characteristic abrupt sound. Normally lasting only minutes, some cases of hiccups have persisted for years. In the worst cases, one treatment is to deliberately damage the phrenic or vagus nerves associated with the diaphragm.*

THE MEDIASTINUM

GROSS ANATOMY | The mediastinum is the space in the chest cavity sandwiched between the two pleural cavities containing the lungs. Importantly, it contains the heart with its outer sack-like covering (pericardial sac), the major vessels going to and from the heart, and almost all of the esophagus. To locate structures more easily, draw an imaginary line horizontally through the mediastinum at the level of the sternal angle (which itself can be felt by running your fingers down the sternum where the upper two parts of the sternum form a joint). The superior mediastinum is above the line, and the great vessels coming off or entering the heart sit in this space. The area below the line is the inferior mediastinum; this is further subdivided into three spaces by the pericardial sac. The contents in front of the pericardial sac (anterior mediastinum) are minimal, with just fat, lymphatics, and the leftovers of the soft, bi-lobed thymus gland located here. The pericardial sac contains the heart and its blood supply (middle mediastinum). The descending aorta, the esophagus (and its nerve supply), the azygos vein, and the lymph-carrying thoracic duct are safely tucked away behind the pericardial sac in the posterior mediastinum.

CLINICAL ANATOMY | Infections from the oral cavity and neck can track into the mediastinum via anatomical pathways existing between these spaces. The potential space adjacent to the pharynx—it is not a real space but one that can become a space if something enters it—is most likely to get infected from a tooth abscess or throat infection, such as those behind the mucous membrane close to the tonsils (quinsy). From here, pus tracks to another potential space behind the pharynx and downward to the potential space around the wrappings of the trachea. This is a direct route into the superior and inferior mediastinum, where the large blood vessels and heart are located. These complicated infections can be life-threatening, as they spread around the body through the bloodstream, or the volume of pus blocks off the airways.

DISSECTION | *The ancient anatomist, Galen, named the ductless lymphoid gland in the mediastinum "thymus," as it resembled a wart-like outgrowth similar to that on a bud of the plant thyme. The gland is largest around puberty and persists into old age, despite shriveling up and becoming virtually defunct. Its contribution to the body's immunological defenses is vital in the early years.*

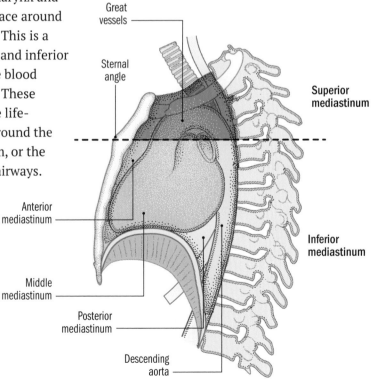

Great vessels

Sternal angle

Superior mediastinum

Anterior mediastinum

Middle mediastinum

Posterior mediastinum

Descending aorta

Inferior mediastinum

THE PERICARDIAL SAC

GROSS ANATOMY | The muscular heart and the roots of the arteries and veins coming off it lie suspended in a conical triple-layered membranous sac (pericardial sac); this provides protection and an inner frictionless surface for an organ that is always beating. The protective outer layer of the sac is thick and strong (fibrous pericardium) and fuses with the tendinous part of the diaphragm below; they move in tandem during respiration. The inner two layers are smooth, moisture-exuding wrappings (similar to those around the lungs). The innermost of these layers covers the heart and cannot be separated from it (visceral pericardium), pushing into every crack and crevice. Like the lung wrappings, it bends back on itself to cover the inside of the tough fibrous pericardium (parietal pericardium). Only a few milliliters of fluid separate these two innermost layers (the space is called the pericardial cavity), which are in constant contact with each other and slide in relation to each other. The tough fibrous layer of the pericardial sac fuses at the top with the great vessels above the heart, making it a closed sac, unlike the pleural layers covering the lungs. It attaches to the sternum and a few ribs at the front via ligaments.

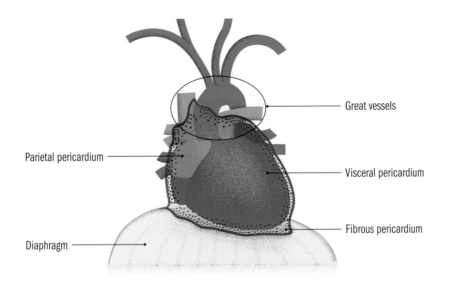

Great vessels

Parietal pericardium

Visceral pericardium

Fibrous pericardium

Diaphragm

CLINICAL ANATOMY | When fluid collects in the pericardial space between the inner two linings around the heart (as a result of a cancer or a bleed following a road traffic accident), the fluid has no way of escaping. This causes the heart to become compressed (tamponade), making the ventricles less able to propel blood around the body. This is a life-threatening medical emergency, for which a long needle is placed through the diaphragm and pericardial sac, into the pericardial space, to drain out the fluid and allow the heart to resume beating properly. In an emergency situation, 3½ fl. oz. (100 ml) of fluid can tamponade the heart enough to cause death.

DISSECTION | *The larger the animal, the slower its heart beats. A canary averages a heart rate of one thousand beats per minute (bpm). Elephants' hearts beat at 25 bpm. In humans, heart rate lessens progressively with growing size up to adolescence. A baby's heart rate is 130 bpm. At rest, an adult's heart rate is 70 bpm, although this can rise significantly on a temporary basis.*

THE HEART

GROSS ANATOMY | The main function of the muscular heart is to ensure a steady supply of oxygen-rich blood is available for every tissue in the body. The heart is a single organ made up of two independent pumps propelling blood simultaneously through two different circuits. One circuit pumps blood to the lungs and the other circuit to body tissues. Blood is received in the upper chambers (atria) and propelled out of the heart through the lower chambers (ventricles). Flow of blood is in one direction only, with four valves (aortic, pulmonary, tricuspid and mitral) at critical points preventing backflow of blood; their coordinated closure in succession causes the characteristic "lub-dub" sound heard through a stethoscope. The size of an individual's heart is approximately the size of their own clenched fist. The bulk of the heart is a tough and layered muscular wall (myocardium), lined on the outside by a thin layer of tissue (pericardium) and on the inside by another layer of tissue (endocardium). A thick wall divides it into right and left sides. The right side of the heart is a low-pressure area and its muscular walls are thin. The left side is a high-pressure area, propelling blood at speed throughout the body, and its muscular wall is much thicker.

CLINICAL ANATOMY | A fetus receives oxygenated blood from its mother via the placenta, so there is no real need to pump blood to the lungs for oxygenation. Instead, two shunts exist in the fetal heart. These close up after birth, leaving reminders of their existence on the adult heart (between the atria and between the great vessels). However, occasionally, one of these shunts stays open. If the opening between the atria remains open, poorly oxygenated blood from the right atrium mixes with oxygen-rich blood in the left atrium. While "holes in the heart" are often not symptomatic, they do put a person at risk of strokes and heart attacks.

DISSECTION | *Two of the four heart valves are held in place to the walls of the ventricles via chordae tendineae (heartstrings), the cord-like tendons or fibrous strings that prevent these valves from being forced into the atria when the ventricles contract. They resemble the strings on a musical instrument. When someone pulls on our heartstrings, our deepest emotions are stirred, figuratively speaking.*

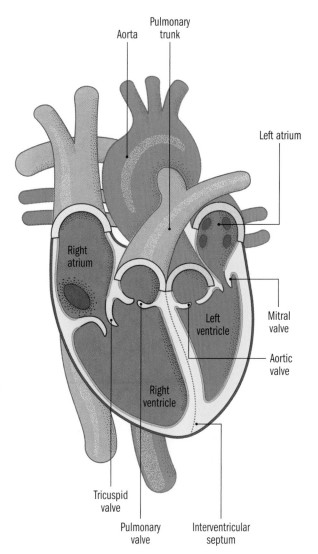

THORACIC ANATOMY 77

CORONARY CIRCULATION

GROSS ANATOMY | Above all, the heart must ensure its own oxygen requirements are met at all times. Even small problems in its own circulation can have devastating effects and reduce its pumping efficiency. The heart's own blood supply comes off the right and left coronary arteries, both of which branch off the wall of the aorta just before blood is propelled to the rest of the body. These arteries supply both of the ventricles and the partition between the ventricles. The right coronary artery supplies the right atrium as well as the heart's own pacemaker (the sinoatrial node generates impulses that control heart rate) situated in the right atrium, although in a third of the population its blood supply may come off the left coronary artery. The left coronary artery is the larger of the two and splits into two large trunks almost immediately. Running forward in a groove between the two ventricles is its most important (and most commonly blocked) artery, the anterior interventricular artery. A second branch travels down a similar groove at the back of the heart. The two coronary arteries circle the heart and terminate at the back, where they have an overlapping blood supply to ensure that all areas of the heart muscle are sufficiently well supplied.

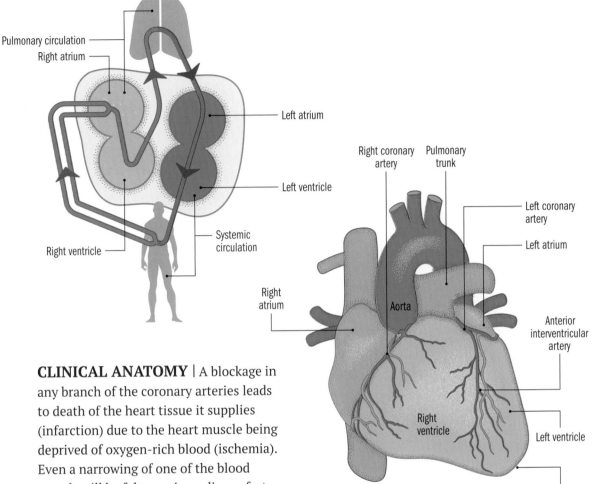

Pulmonary circulation

Right atrium

Left atrium

Left ventricle

Right ventricle

Systemic circulation

Right coronary artery

Pulmonary trunk

Left coronary artery

Left atrium

Right atrium

Aorta

Anterior interventricular artery

Right ventricle

Left ventricle

Apex

CLINICAL ANATOMY | A blockage in any branch of the coronary arteries leads to death of the heart tissue it supplies (infarction) due to the heart muscle being deprived of oxygen-rich blood (ischemia). Even a narrowing of one of the blood vessels will be felt as pain or discomfort (angina pectoris). Over time, fatty plaques build up in the arteries and narrow the coronary vessels. When a plaque becomes big enough to slow down or block an artery altogether, it leads to a heart attack (myocardial infarction), where a section of heart muscle dies. Unless unblocked quickly, heart failure or uncoordinated and uncontrolled contraction of the ventricles may follow, both of which can be fatal.

DISSECTION | *"Corona" comes from Latin and means crown. It is used in relation to several anatomical structures in the body. The anatomical use of the word "coronary" began in the 1600s, when the structure "encircling the heart like a crown" (i.e., the arteries) was first referred to. The term "coronary artery" was coined in the 1700s and has been in use ever since.*

THE AORTA

GROSS ANATOMY | As wide as a garden hose and shaped like candy cane, the aorta is the largest artery in the body. It distributes oxygen-rich blood straight from the heart to every cell, organ, and structure in the body via a rich network of branching vessels. The artery begins at the base of the left ventricle of the heart, and blood is ejected out at high pressure via a three-part valve (aortic) that, once ejected, prevents blood from flowing backward. Two small openings in the wall are the origin of the heart's own arterial supply, the coronary arteries. Blood is propelled upward via the short ascending aorta, then arches backward (arch of the aorta) and descends into the back of the chest cavity as the descending aorta, tucked away behind the heart. Three large branches off the arch of the aorta propel blood upward into the head, neck, and upper limbs. The descending aorta gives off numerous branches within the chest (thoracic aorta) and travels through an opening behind the diaphragm (aortic hiatus) to supply the abdomen (abdominal aorta). Bundles of arteries branch off to supply the abdominal organs, often giving a dual and overlapping supply. Above the pelvis, the aorta splits into the iliac arteries, which supply everything below this region.

CLINICAL ANATOMY | The wall of the aorta is made of three muscular layers. Abnormal widening (aneurysm) may happen anywhere along its course as a result of aging, untreated high blood pressure, or rare diseases. A tear may occur in the inner of the three linings, with blood leaking into the middle layer and creating a false passageway for blood flow. The arterial wall is further weakened as blood is diverted into this cul-de-sac, which becomes bigger and bigger. A split (aortic dissection) can spread and involve structures closer to the heart or the pericardial sac. All of these conditions require urgent surgical intervention, as a rupture can be deadly.

DISSECTION | *A branch of the left vagus nerve, on its way through the neck and chest to the abdomen, loops underneath the arch of the aorta to travel upward to supply the larynx. This is an anatomical feature humans share with several other vertebrates. In giraffes, the nerve covers a distance of 15 ft (5 m); in humans this is less than 4 in. (10 cm).*

Closed aortic valve

Open aortic valve

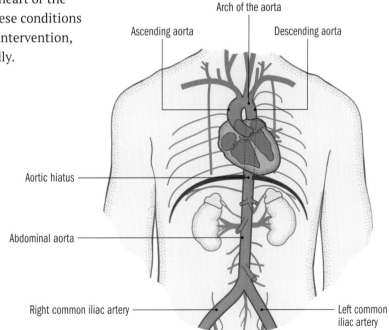

Arch of the aorta

Ascending aorta

Descending aorta

Aortic hiatus

Abdominal aorta

Right common iliac artery

Left common iliac artery

THE VENA CAVAE

GROSS ANATOMY | Oxygen-depleted blood returns
to the right side of the heart via two main veins above
and below the right atrium into which they empty their
contents. The superior vena cava (SVC) receives blood
from the top end of the body from structures above the
diaphragm. It is formed by two large veins (brachiocephalic)
coming together in the midline of the upper part of the
chest. On average, the SVC measures around $2^3/4$ in. (7 cm).
As these are low-pressure vessels, the SVC has no valves
preventing backflow of blood and blood flows passively
into the right atrium of the heart, to be circulated into
the lungs for re-oxygenation. Tissues below the diaphragm
drain into the inferior vena cava (IVC), which is formed
on the back wall of the abdomen by two major (iliac) veins
draining the lower limbs and pelvis coming together.
The majority of the IVC is within the abdomen, and the
thoracic part is very short and partly embedded within
the pericardial sac surrounding the heart. The IVC travels
upward through an opening on the right side of the
diaphragm to reach the thorax (caval opening).

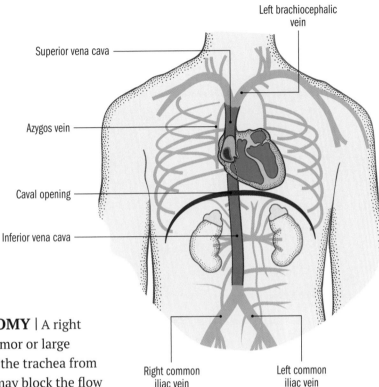

Left brachiocephalic vein

Superior vena cava

Azygos vein

Caval opening

Inferior vena cava

Right common iliac vein

Left common iliac vein

CLINICAL ANATOMY | A right superior lobe lung tumor or large lymph nodes around the trachea from a metastatic cancer may block the flow of blood through the SVC into the right atrium. This superior vena caval obstruction manifests as headaches, engorged vessels in the face and neck, and facial swelling. It requires urgent treatment to relieve the symptoms, usually in the form of a stent placed inside the SVC to keep it open, or radiotherapy to shrink structures. If the IVC becomes obstructed, alternative (and quite convoluted) re-routing of the blood can take place to ensure the low-oxygen blood in the abdomen returns to the heart.

DISSECTION | *Most anatomical structures are symmetrical, so when an asymmetrical single vein was discovered by anatomists in the distant past, they named it "azygos" after the Greek word meaning "unpaired." The vein overlies the vertebral column, more to the right than the left. It drains the majority of the back wall of the thorax and empties into the superior vena cava.*

ESOPHAGUS

GROSS ANATOMY | The esophagus (food pipe) is a 10 in. (25 cm) long, fairly straight muscular tube that propels chewed food and fluids from the pharynx into the stomach. When relaxed, it is flat. In the neck, the flattened esophagus sits directly behind the trachea, into which it bulges when a bolus of food passes through it. In the thorax, it is behind the pericardial sac and in front of the bony spine projecting into the chest. The lymph-containing thoracic duct and azygos vein run behind it, and the vagus nerves on either side, along its course, innervate it. Anatomically, the esophagus is composed of four different layers, one of which is formed of both circular and longitudinal muscles for propelling food downward. It has a rich blood supply and an extensive network of veins encompassing it, vital for all parts of the digestive tract. The esophagus has gatekeepers (sphincters) at both ends which keep it sealed off and prevent backflow of contents. The lower sphincter is not anatomically a true sphincter (a circular muscle that constricts, resembling strangulation). Instead, as the esophagus makes its way into the abdominal cavity, fibres from the diaphragm slings around it at the esophageal hiatus and constricts it on inspiration. This stops contents from the acidic stomach from going backward but it can be relaxed to allow belching or vomiting.

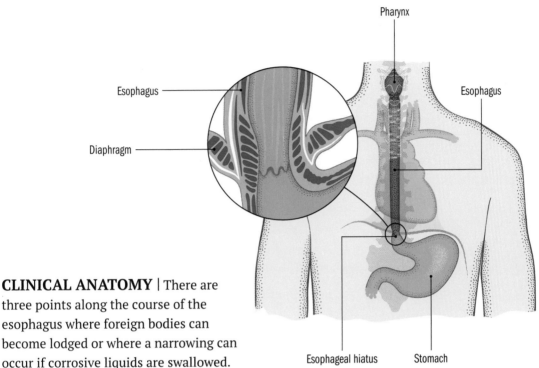

Pharynx

Esophagus

Diaphragm

Esophagus

Esophageal hiatus   Stomach

CLINICAL ANATOMY | There are three points along the course of the esophagus where foreign bodies can become lodged or where a narrowing can occur if corrosive liquids are swallowed. Measured from the upper front teeth at 6, 10, and 16 in. (15, 25, and 41 cm), they correspond to the start of the esophagus, where the left bronchus crosses it and where it enters the abdomen through the diaphragm. Problems with the liver can mean blood is redirected to the heart via smaller, lesser-traveled veins around the esophagus, which weaken and bulge and can bleed copiously. Violent vomiting can tear the inner esophageal lining. Both can result in bloody vomit or sweet-smelling, black tarry stools from digested old blood.

DISSECTION | *Boluses of chewed matter are pushed along the digestive tract through peristalsis, the continuous involuntary wave-like motion of muscles in these organs. In the esophagus, the waves travel the whole length of the tube to push contents into the stomach. It takes about nine seconds for a wave to travel from the start of the esophagus to the stomach.*

THORACIC DUCT

GROSS ANATOMY | Lymph from the lower limbs and abdomen travels via the thoracic duct to be returned to the venous circulation in the lower part of the neck and upper thorax. The approximately 15–17¾ in. (38–45 cm) long duct starts at a collecting area (cisterna chyli) lying against the spine in the abdominal cavity and travels upward sandwiched between the azygos vein and the descending aorta behind the diaphragm. The esophagus overlies it. At its origin in the abdomen, it is less than ¼ in. (6 mm) diameter, but narrows even further along its course upward. In half the population, it widens slightly before depositing its contents into the junction between two large veins in the neck (subclavian and internal jugular veins). The duct is a bit curvy and narrows every now and then along its course. Inside, it has valves at places where it is exposed to pressure, as the fluid needs to travel upward against gravity. To ensure lymph empties into the venous circulation, and venous blood does not regurgitate into the thoracic duct, a valve with two cusps (bicuspid) faces into the vein at the exit point for lymph. After death, the two substances mix freely as the valve does not work anymore.

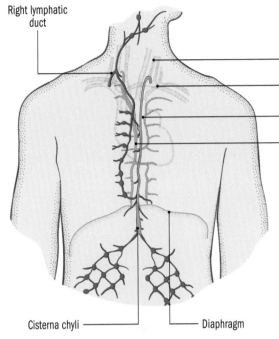

Right lymphatic duct

Internal jugula vein

Subclavian vein

Thoracic duct

Azygos vein

Cisterna chyli

Diaphragm

DISSECTION | *Lymph in most body tissues is clear and colorless, as it is a form of blood plasma (interstitial fluid). Chyle is a specialized milky-white substance from the small intestine, mostly made of small fat globules and lymphatic fluid. There are no lymphatics from the thymus gland, the cornea of the eye, bone marrow, or the central and peripheral nervous systems.*

CLINICAL ANATOMY | The lymphatic system can become blocked as a result of scarring of the inner duct wall, metastatic cancer, or infection from a parasitic roundworm (filariasis or elephantiasis). The lower limbs and scrotum become swollen and engorged. This can lead to blockage in the main ducts and milky-white chyle can leak into the spaces in the wrappings around the lungs and abdomen, complicating the disease even further by causing breathing difficulties, even leading to death. Occasionally, fractures to the thoracic vertebra can tear the duct or it can become damaged during esophageal surgery. Both of these can lead to chyle filling the pleural space surrounding the lungs (chylothorax).

"Those who have dissected or inspected many [bodies] have at least learnt to doubt; while others who are ignorant of anatomy and do not take the trouble to attend it are in no doubt at all."

GIOVANNI BATTISTA MORGAGNI
DE SEDIBUS ET CAUSIS MORBORUM (1761)

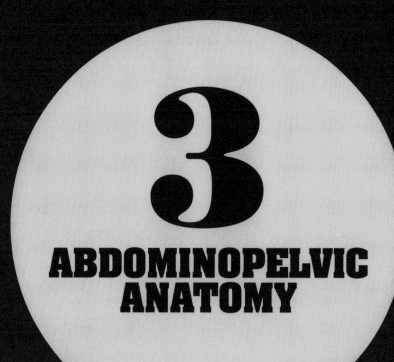

3

ABDOMINOPELVIC ANATOMY

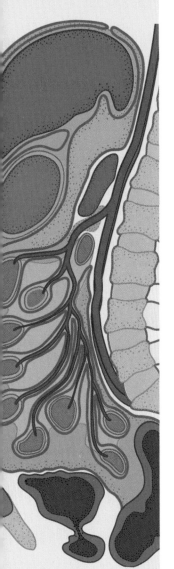

DIGESTION & EXCRETION

The digestive and urinary systems constantly communicate to maintain a balance between absorbing nutrients and expelling waste. Both systems are regulated through input from the neuroendocrine system. The cardiovascular system is also linked on multiple levels with kidney and bowel function, and blood volume is tightly regulated by the kidneys, ensuring blood pressure is not too high or too low. The kidneys also work together with the respiratory system to make sure the concentration of blood is maintained within a tight range (not too acidic and not too alkaline). Each system on its own has a purpose: the digestive system to digest food and provide the body with nutrients, and the urinary system to remove toxins and fluids our body does not need. They rely on each other and their constant interplay keeps us well. Problems in one system can affect the other. If, for instance, the kidneys are malfunctioning, toxins arriving from the digestive system cannot be removed and recirculate through the body, leading to a buildup of toxins with subsequent harmful effects.

Our digestive journey

Our digestive tract extends from our mouth to our anus, connected by a series of interconnected tubes that are adapted for specialized functions along its course. Through this tract we take in food, break it down into molecules that can be used by the tissues and cells in our body, absorb nutrients we need, and get rid of waste. Several organs other than the gut partake in this process—the gallbladder, liver, salivary glands, and pancreas are all part of the digestive system. We take in food through our mouths, where our teeth grind it, our tongue manipulates it and moves it around our mouth, and saliva from three major salivary glands moisten this ingested material, all contributing to making it into a ball (bolus) that can slide down our throat (pharynx) into the food pipe (esophagus). Taste buds littered all over the surface of the tongue help us to quickly assess whether to enjoy the food or spit it out (if the taste suggests it may be harmful to us). Acceptable food is swallowed as the tongue pushes the bolus up against the hard palate and past the soft palate into the pharynx (throat), from where it is propelled downward by waves of contractions (peristalsis) via the esophagus into the stomach.

When food arrives in the stomach, it undergoes an "acid attack" to chemically break it down. It is also churned and then propelled into the start of the small intestine (duodenum). The duodenum is connected to the pancreas and liver by a series of tubes that bring in pancreatic enzymes and bile, all tailored to dissolve and digest carbohydrates, protein, fiber, and fat. The small intestine (duodenum, jejunum, and ileum) is coiled up in the middle of the abdomen and covered on its inner walls by tiny projections (microvilli). The microvilli increase the surface area of the small intestine for a very rich blood supply, and are connected to capillaries. It is through this interface that all digestible nutrients pass from the small intestine into the bloodstream and, from there, into the liver to be further processed. The majority of absorption happens in the small intestine, though alcohol (and some types of drugs) get partly absorbed from the stomach. Anything your intestines cannot break down (fiber, for instance), as well as bile and millions of bacteria, passes through the M-shaped large intestine and into the rectum for storage until they can be passed out.

Journey from blood to urine

Our kidneys excete excess water and the end products of metabolism, which is essential to maintaining the correct concentration of substances within the body, and the correct water and electrolyte balance in tissue fluids. The nutrients from the watery mass in our small intestine make their way into our bloodstream (circulatory system) and from there into the liver via the portal system (all blood from the digestive system goes to the liver first for processing), and then via the circulatory system into the urinary system (kidney, ureters, bladder, and urethra). All blood passes through our kidneys, which sit high up on the back wall of the abdomen. The kidneys filter out waste products (particularly urea) from the blood as it passes through an intricate meshwork within them. The functional unit of the kidney is the nephron, and each kidney is composed of a million nephrons. Multiple sites along the course of a nephron reabsorb substances the body needs. Waste products and excess fluid are formed into urine. Urine then flows down ureters to the bladder, an expandible muscular bag that stores and then releases the urine through the urethra when it is full.

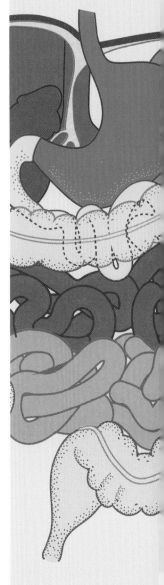

ANTEROLATERAL ABDOMINAL WALL

GROSS ANATOMY | Internal organs in the abdominal area have limited protection from the external world, as there is no underlying bony framework to shield them from damage at the front and the sides. Their protection is from a layered muscular wall, padded on the exterior by skin and fat and on the interior by sheets of thin membrane, the muscles of which can tense to form a rigid protective wall against any external onslaught. Three sheets of muscles—external oblique, internal oblique, and transversalis abdominis—emerge from the sides, and their fibers travel mainly in opposing directions, crisscrossing each other to provide strength to the wall. Their tendons flatten and merge just before they encounter two vertical strips of muscle on either side of the midline, the rectus abdominis or "six-pack" muscles in a toned abdomen. These thin sheets of muscles forming the anterolateral abdominal wall move the trunk, enabling twisting, sideways, and forward movement of the spine. They recoil passively when breathing out, pulling the ribcage down. The muscles help maintain posture and support the spine in the upright position when sitting or standing. Tensing them flattens the abdomen and raises the pressure inside the abdomen, allowing emptying of the bladder, defecation, vomiting, coughing, singing, childbirth, lifting heavy objects, and forcing air out of the lungs.

CLINICAL ANATOMY | The muscular abdominal wall is the access point for surgical procedures in the abdomen and pelvis; its anatomy is of great relevance to the surgeon aiming to avoid unnecessary blood loss and improve postoperative outcomes. Rapid access to the abdomen can be gained through a vertical incision in the midline, which is a virtually bloodless white line (linea alba) where the flattened tendons (aponeurosis) of the abdominal wall muscles meet below skin and fat. Incisions made a few inches parallel to the linea alba, and away from the midline rectus abdominis muscle, avoid damaging the muscles and the large blood vessels that run directly under it, and form an important anastomosis just below the umbilicus (bellybutton).

DISSECTION | *Small, paired triangular abdominal wall muscles may overlie the lower end of the rectus abdominis muscles, although they vary in size, shape, and number. These pyramidalis muscles, when present, tense the midline of the abdominal wall, the linea alba, onto which they attach. It is still unclear what functional significance this has in humans.*

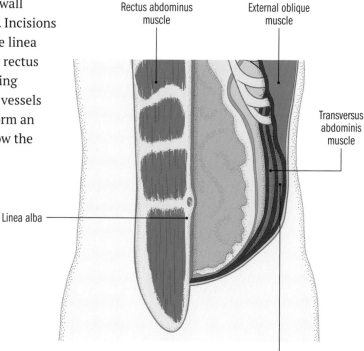

Rectus abdominus muscle

External oblique muscle

Transversus abdominis muscle

Linea alba

Internal oblique muscle

INGUINAL CANAL

GROSS ANATOMY | The transition zone where the trunk meets the thigh is an area known as the inguinal region. The lower part of the muscular anterolateral abdominal wall ends here, terminating in a ligament formed by tendinous folds of one of the broad flat muscles that comprises it. Overlying this inguinal ligament, on either side, is a $1\frac{1}{2}$ in. (4 cm) long cylindrical passageway tunneling through the layers of the abdominal wall. The inguinal canal allows structures to travel and communicate between the pelvis and outside the abdominal wall. Structures pass through the canal via small slit-like openings on the inside (deep inguinal ring) and outside (superficial inguinal ring) of the canal walls. The areas in front and behind these openings are reinforced to retain strength in the wall. In males, the spermatic cord carries numerous structures to and from the scrotum (sperm travels through the vas deferens from the scrotum into the urethra of the penis inside the pelvic region). In females, a ligament originating on the uterus (womb) runs into the inguinal canal and makes its way to the external genitalia, tethering the uterus to a position where it is facing forward.

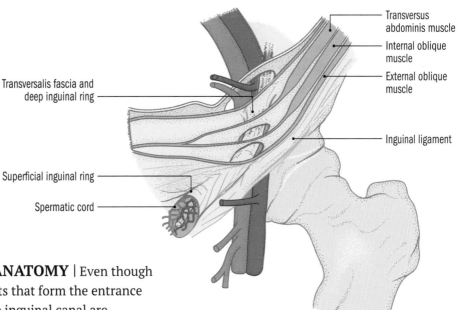

Transversus abdominis muscle

Internal oblique muscle

External oblique muscle

Transversalis fascia and deep inguinal ring

Inguinal ligament

Superficial inguinal ring

Spermatic cord

CLINICAL ANATOMY | Even though both of the slits that form the entrance and exit to the inguinal canal are reinforced, they still represent a breach in the muscular wall that can widen as the wall weakens. Pressure on the wall from within can weaken the openings, and abdominal contents (intestines) can push through (hernia). Direct hernias push contents through the superficial ring closest to the midline in the groin area, while indirect hernias travel through the deep ring, along the inguinal canal and exit through the superficial ring. Usually, hernias can be pushed back but their contents can sometimes become trapped and their blood supply cut off.

DISSECTION | *Inguinal stems from the Latin word "inguen," the oblique depression between the abdomen and the thighs. To ancient Romans, inguen also signified one's "privates" (sexual organs). The Middle English word "groin" (grynde) meant an abyss, hollow, or depression (in the ground). Crotch, used informally, is from Old French, meaning pitchfork or shepherd's crook, i.e., the region where the body forks.*

ABDOMINAL & PERITONEAL CAVITY

GROSS ANATOMY | The abdominal cavity is the largest hollow space within the body. It houses the liver, pancreas, kidneys, adrenal glands, the spleen, and the majority of the digestive tract. At the back, it is supported by the vertebral column, the lower part of the ribcage, the pelvic bone, and sturdy back muscles. The diaphragm separates it from the thoracic cavity above. Although there is no clear anatomical boundary between the abdominal cavity and the pelvis below, the pelvic contents are lined by a thin layer of membrane (peritoneum). Peritoneum surrounds most of the inside of the abdominal cavity (parietal peritoneum) and folds back on itself to cover every organ within (visceral peritoneum). A small amount of fluid lubricates the vacuum-like potential space between the two layers (peritoneal cavity), which stretches around the abdomen to envelop structures within. As the organs are packed together closely, the peritoneum provides a sliding surface that minimizes friction. It also holds organs in place and connects some organs with others. Two of its folds are of special relevance: the four-layered greater omentum hangs off the stomach, like an apron, and the small bowel mesentery, which gathers the long sausage-like loops of small intestine together and attaches them to the back of the abdomen. Blood vessels, lymphatics, and nerve supply run within these folds.

CLINICAL ANATOMY | Inflammation of the peritoneal cavity occurs when bacteria enter the sterile potential space, commonly from a rupture in the wall of the intestines or appendix. In life, intestines are filled with bacteria. The resulting widespread inflammation of the peritoneal cavity (peritonitis) is a surgical emergency. It requires immediate treatment with antibiotics and, if necessary, repairing the rupture. Peritoneal folds connecting structures contain blood vessels. Those running inside the edge of a fold known as the lesser omentum (between the stomach and liver) can be accessed if there is a liver bleed during an operation. In a procedure known as the Pringle maneuver, the fold is tightly squeezed for a short period to tackle the source of the blood loss.

DISSECTION | *Dangling down like a fatty apron to cover the front of the internal organs, and known as the "policeman of the abdomen," the highly mobile greater omentum can migrate and wrap around damaged or infected organs, containing infection and preventing spread. It provides an excellent place to store fat. Men are particularly susceptible to storage here (known as visceral fat), as evidenced by pronounced "beer bellies."*

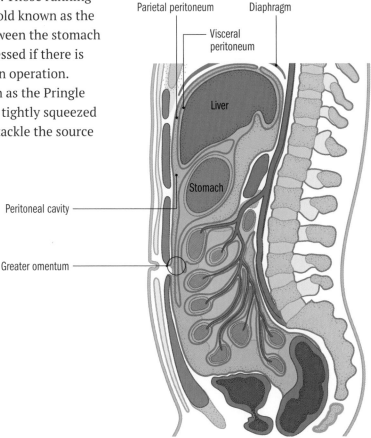

Parietal peritoneum Diaphragm

Visceral peritoneum

Liver

Stomach

Peritoneal cavity

Greater omentum

Sagittal view of abdominal cavity

STOMACH

GROSS ANATOMY | The stomach is a somewhat J-shaped expandable muscular bag between the esophagus and the small intestine. Sat at the front of the upper left side of the abdominal cavity, it receives food from the esophagus. The process of digestion starts in the stomach. Food is exposed to a cocktail of powerful acids (hydrochloric acid) secreted from cells lining the main body of the stomach, and the contents are churned by the triple-layered muscular wall and then propelled into the small intestine. The stomach's entrance (lower esophageal sphincter) and exit points (pyloric sphincter) are guarded by sphincters preventing back flow. They allow boluses of food to pass through at set intervals. Close to its exit point, the cells produce a more alkaline secretion, neutralizing the contents somewhat before they are emptied via the pylorus into the duodenum, the start of the small intestine. The inner lining of the stomach is folded into numerous wrinkles (rugae). As food fills the stomach, the wrinkles smooth out and the stomach expands. The shape, position, and size of the stomach varies greatly, depending often on whether it is empty or full. It tends to be elongated in tall and thin people but, in short and stout individuals, may lie high up and across the upper abdominal area.

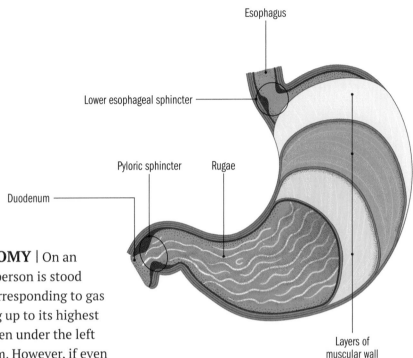

Esophagus

Lower esophageal sphincter

Pyloric sphincter

Rugae

Duodenum

Layers of
muscular wall

CLINICAL ANATOMY | On an X-ray taken when a person is stood up, an air bubble (corresponding to gas in the stomach rising up to its highest point) is normally seen under the left side of the diaphragm. However, if even a small rupture occurs in the digestive tract, air (and bacteria-laden bowel contents) escapes into the sterile and vacuum-like peritoneal cavity. This is visible as a subtle line (air under the diaphragm) on an X-ray, requiring urgent surgical management. Stomach ulcers can sometimes erode through the muscular wall. If erosion occurs through the back wall, this can, very rarely, damage a large artery (splenic) behind it, leading to a catastrophic internal bleed.

DISSECTION | *The pylorus has a thickened circular band of muscle that acts like the gatekeeper between the stomach and the small intestine. The Greek word means "band" or "anything that binds tight," and was first used in the anatomical sense by the ancient anatomist, Galen. More than 60 sphincters are scattered around the human body, some minuscule and visible only under a microscope.*

SMALL & LARGE INTESTINE

GROSS ANATOMY | Measuring about 23 feet (7 m) long, the three-part small intestine (duodenum, jejunum, and ileum) is the longest (though not the widest) section of the digestive tract. Its sausage-like coils are held together by peritoneal membrane, through which travels a wide network of blood vessels. Wavelike contractions (peristalsis) running through the entire digestive tract propel partly digested food from the stomach through the pyloric sphincter into the horseshoe-shaped duodenum, where pancreatic juices, bile, and sodium bicarbonate are added to the mix. These break down proteins, fat, and carbohydrates, and further neutralize the acidic contents. The majority of digested contents are taken up into the bloodstream through the lining of the small intestine, which is covered in microscopic finger-like velveteen protrusions which increase its surface area. These microvilli contain minute blood vessels (capillaries) that allow nutrients to enter the bloodstream. Food that cannot be digested passes through to the large intestine, where water is removed, hardening the contents to form feces. Peristaltic waves push fecal matter through the 5 ft (1.5 m) long colon (cecum and ascending, transverse, descending, sigmoid colon) and the rectum, from where it is excreted via the anus. The rectum and anus have sphincters that allow feces to pass through.

CLINICAL ANATOMY | The cecum is the pouch-like start of the large intestine into which the contents of the small intestine empty. Dangling from it, often behind it, is a wormlike 3–4 in. (8–10 cm) long hollow tube, commonly known as the appendix. While its function is still heavily debated, the appendix's narrow opening into the cecum is prone to blockage (usually by hardened fecal matter); its mucous secretions cannot then be expelled and cause the organ to swell and enlarge. With its blood supply compromised, the tissue dies (necrosis). Filled with multiplying bacteria, a burst appendix can lead to these contents entering the sterile peritoneal cavity, causing severe infection (peritonitis).

DISSECTION | *Around 15 minutes to an hour after eating or drinking, increased electrical activity from stretching of the stomach and small intestine triggers a physiological reflex known as the gastrocolic reflex. The resulting giant migrating contractions come in huge tidal waves, pushing contents out to make space quickly. This is experienced as the strong urge to defecate.*

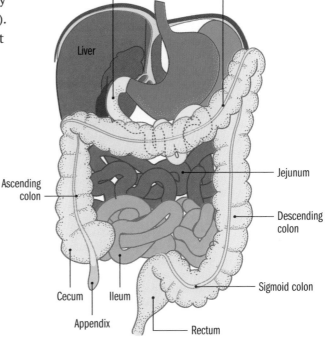

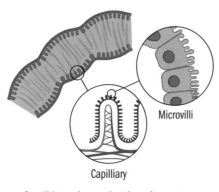

Microvilli

Capilliary

Small intestine under the microscope

LIVER, GALLBLADDER & PANCREAS

GROSS ANATOMY | The liver is the largest internal organ and fulfills numerous vital functions. A large vein (portal vein) carries nutrient-rich blood from the small intestine to the liver for processing. Specialized liver cells (hepatic cells) make up around 60 percent of liver tissue, converting most of the nutrients into forms the body can use. The liver converts and stores sugars, regulating blood sugar levels. It breaks down fats, produces cholesterol and detoxifies the body after drug and alcohol consumption. Importantly, it also produces blood-clotting factors that stop profuse bleeding after injury. A greenish-yellow, mildly acidic fluid (bile), which breaks down fats in the food, is produced by the liver but stored and concentrated (water is removed) in the gallbladder. The gallbladder is on the underside of the liver. Around a liter of bile is produced daily. Following a fatty meal, the gallbladder contracts and squeezes bile into the duodenum to break down the fat. The pear-shaped and light (only 2¾ oz /80 g) pancreas, sitting snugly in the C-shape of the duodenum and behind the stomach, is a gland that secretes digestive juices into the duodenum to break down carbohydrates, protein, and fats and neutralize stomach acid. It also secretes blood-sugar regulators, insulin, and glucagon directly into the blood.

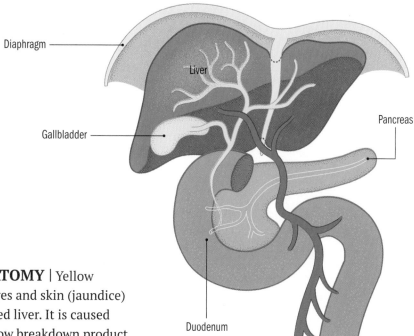

Diaphragm

Liver

Gallbladder

Pancreas

Duodenum

CLINICAL ANATOMY | Yellow discoloration of eyes and skin (jaundice) indicates a damaged liver. It is caused by bilirubin, a yellow breakdown product of red blood cells, building up in the blood. Life is not possible without a liver. However, if part of it is removed, its tissue can regenerate over weeks. When insulin production in pancreatic cells is disrupted, high levels of sugar circulate in the blood, damaging vessels. Sugar cannot be taken up and stored in the liver. Without insulin replacement, this results in unquenchable thirst as sugar causes fluids to be removed from the body, as well as copious amounts of urine and a breakdown of fat and tissue. This is diabetes mellitus.

DISSECTION | *Ancient Indian surgeon, Sushruta, (5th century BCE) observed ants gathering around the sickly sweet smelling and tasting urine of diabetics. It was only in the 17th century, in the Western world, that a doctor and neuroanatomist from Oxford, Thomas Willis, first coined the term "diabetes mellitus" from the Greek for "siphon" (a bent tube)—to pass through— and the Latin word meaning "honeyed."*

SPLEEN

GROSS ANATOMY | The soft and purple fist-shaped spleen sits in the upper left-hand corner of the abdominal cavity, wedged between the diaphragm and the stomach, and well-protected by the ribcage surrounding it. It is encased in a thick capsule. Weighing $5\frac{1}{4}$ oz. (150 g) in an average adult, the organ is very well supplied with blood and its mass is made up of two types of tissue that intermingle but have different functions. Red pulp acts as a blood filter, sifting out bacteria, viruses, and other debris. It also specializes in destroying tired red blood cells. The life span of a red blood cell is around 120 days, after which the spleen breaks it down and the breakdown products are transported elsewhere for recycling or removal. White pulp is like a massive lymph node, and produces lymphocytes, a type of white blood cell that helps fight off infection by initiating an immune response when a foreign substance (antigen) is encountered. As blood flows through the spleen, white blood cells remove foreign invaders, keeping blood clear of infectious agents. In the developing fetus the spleen produces red and white blood cells but loses this ability just before birth, after which the bone marrow takes over this task.

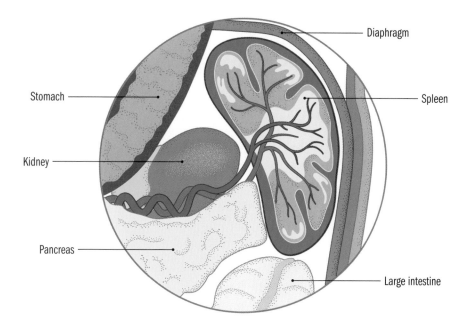

Diaphragm

Stomach

Spleen

Kidney

Pancreas

Large intestine

CLINICAL ANATOMY | Even though the spleen is well protected, in accidents its capsule can be damaged or the vessels supplying it can tear. Torrential blood loss may follow. While it is possible to live without a spleen, salvaging as much as possible is safest. A partly removed spleen may regenerate without any evident loss of function. A fairly normal life is still possible even if the spleen is completely removed, but problems may occur when the body encounters certain encapsulated bacteria and an overwhelming severe blood infection (sepsis) can follow. To prevent this, those who have had their spleen removed require a low dose of antibiotics for the rest of their life.

DISSECTION | *The shape of the spleen ranges from a domed tetrahedron to a slightly curved wedge, determined by the structures pressing on it during fetal development. Its size and weight are determined by age and, partly, by the amount of blood within it. Curiously, when the spleen enlarges (for instance, as a consequence of infection or liver disease), it travels diagonally across the abdomen, past the belly-button, and may be felt in the lower-left hand corner.*

POSTERIOR ABDOMINAL WALL

GROSS ANATOMY | The posterior abdominal wall is not easily defined. It is the area at the back of the abdominal cavity that is not covered by the anterolateral abdominal wall. Behind it, and closer to the skin surface, are the back muscles covered by skin, fat, and fascia. The muscles forming this back wall move the lower limb or the spinal column and run inside the abdominal cavity. The bony pelvis and the vertebral column provide protection for the numerous organs overlying the muscles. The kidneys, adrenal glands, pancreas, duodenum, large intestine, ureters, nerves, and the aorta and inferior vena cava all reside in this tight space. These structures sit behind a double-layered membranous sac (peritoneal sac) enclosing the other abdominal organs. This is the retroperitoneal space ("retro" in this case meaning "behind"). The organs are sandwiched between the membranous sac in front of them and the muscular wall behind them. The region can be further divided into several spaces close to the kidneys and ureter based on layers of fascia splitting around these structures. Because the retroperitoneal space is so tightly packed, with little movement inside, small amounts of pus, blood, and fluid can be contained within them without spreading afar and causing widespread damage.

CLINICAL ANATOMY | The psoas major muscles (which bend the hip) originate from either side of the spinal column inside the abdomen, descend past the pelvis, and attach to the thigh bone (femur). The fascia covering them follows them into the upper thigh. Infections from the retroperitoneal area (or around the spine) that break through this fascial covering may track down with the muscle into the thigh, from where spread into large blood vessels can lead to widespread blood infection. Even large bleeds in the retroperitoneal region may go undetected. Bruising appearing in the flanks (between last rib and hip) may indicate severe inflammation of the pancreas or a bleed into the retroperitoneal area.

DISSECTION | *Absent in around 40 percent of the population, a thin long slip of a muscle—psoas minor—overlies its chunky, muscular neighbor, the psoas major muscle. When present, it may assist in bending the trunk. Ancient Greek anatomists described each psoas muscle as being in the shape of a fox's tail. This is particularly the case for the psoas major muscles.*

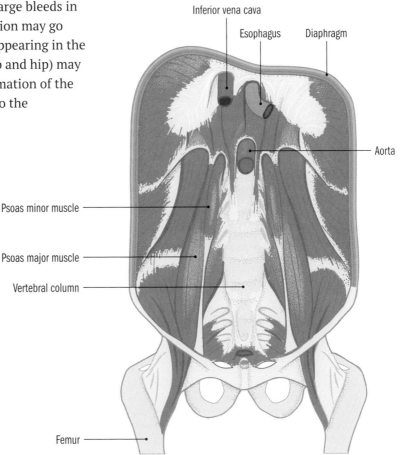

Inferior vena cava

Esophagus

Diaphragm

Aorta

Psoas minor muscle

Psoas major muscle

Vertebral column

Femur

URINARY SYSTEM

GROSS ANATOMY | Shaped like the beans that resemble them and bear their name are the two reddish-brown kidneys sitting high up at the back of the abdominal cavity overlying the posterior abdominal wall muscles. Well protected by the bottom of the ribcage and located on either side of the vertebral column, the kidneys are the start of the urinary system. They produce urine from waste products and excess water in the body, filter the blood, and help maintain stable blood pressure. They keep the delicate composition of blood at a constant, by monitoring and balancing water, pH, and salt levels. The bladder is an expandable hollow muscular reservoir sat behind the pelvic bone for storing urine. Urine is transported to the bladder via ureters, two long narrow tubes running in the retroperitoneal region. The bladder stretches as it fills up with urine and can hold around 10 fl. oz. (300 ml) quite comfortably. The signal to empty the bladder is conveyed in the need to urinate. This can become very painful if the bladder is not emptied. Urine leaves the body via the urethra, a tube of varying lengths in men (around 8 in/ 20 cm long with multiple bends) and women (around 1½–2 in/4–5 cm) long, short, and very slightly curved). A sphincter keeps the junction between bladder and urethra closed until it is acceptable to urinate. The bladder then contracts, and urine is forced out through the urethral opening.

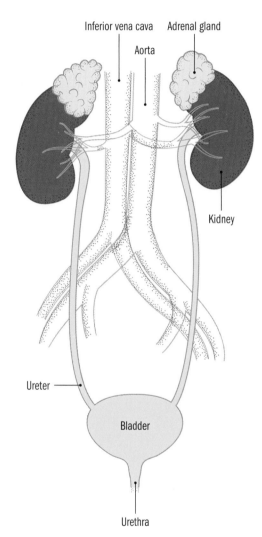

Inferior vena cava

Aorta

Adrenal gland

Kidney

Ureter

Bladder

Urethra

CLINICAL ANATOMY | The kidneys work nonstop to filter around 320 pt. (150 l) of blood every 24 hours. Every minute, 25 percent of circulating blood makes its way through them, producing more than 3 pt. (about 1.5 l) of urine per day. As efficient as they are, problems with the kidneys are difficult to detect in the early stages, and people can live without any clear symptoms for a long time. Just one functioning kidney is enough for survival. When around 90 percent of kidney function is lost, though, a dialysis machine (which replicates the blood-cleaning function of kidneys) is necessary for survival. If dialysis fails, organ donation is the only other option to ensure survival.

DISSECTION | *The kidneys are thought to have several pacemaker sites in the smooth muscle cells of their walls. These trigger wave-like contractions (peristaltic waves) running through the ureters, making wriggly worm-like movements that push urine toward the bladder. Peristaltic waves continue even after partial removal of the kidney, allowing normal flow of urine out of the affected kidney.*

BONY PELVIS

GROSS ANATOMY | The bony pelvis is a ring of bones
in the shape of a basin at the lower end of the trunk.
It supports the weight of the body and cradles some
reproductive and abdominal organs. It is formed by
the fusion of paired hip bones (a further fusion of three
smaller bones: ischium, ilium, and pubis) and the sacrum.
These three smaller bones join together in puberty in
a deep depression known as the acetabulum, forming a
ball-and-socket joint with the head of the femur (thigh
bone) that connects the hip with the thigh. The ilium fans
upward in a wing-like expansion where one can easily rest
one's hands. The ischium is the bony protrusion on which
one sits; strong hamstring muscles that move the thigh
attach to it. The pubic bones fuse at the front via the
symphysis pubis, a flexible cartilage bridge just in front
of the bladder. The genitalia attach to an arch formed by
the pubic bones (pubic arch). The greatest sex differences
in the human skeleton are in the bony pelvis, as the female
pelvis must accommodate a baby passing through during
childbirth. To achieve this, the female pelvis is wider and
shallower, with an oval-shaped rim where hip bones
and sacrum meet, and a wider angle at the pubic arch.
In a male, this pelvic brim is heart-shaped and narrower.

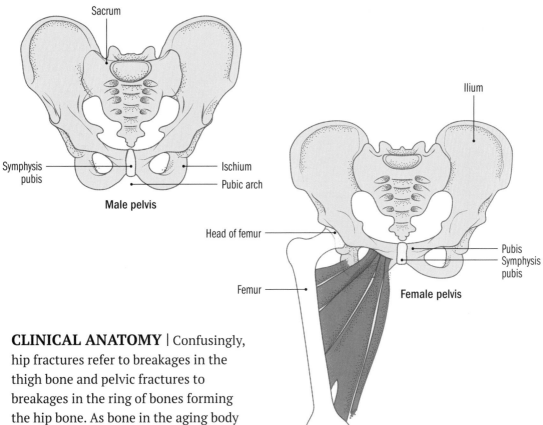

Sacrum

Symphysis
pubis

Ischium

Pubic arch

Male pelvis

Ilium

Head of femur

Pubis
Symphysis
pubis

Female pelvis

Femur

CLINICAL ANATOMY | Confusingly, hip fractures refer to breakages in the thigh bone and pelvic fractures to breakages in the ring of bones forming the hip bone. As bone in the aging body weakens and becomes more brittle, a simple fall may cause a pelvic fracture in the elderly. The pelvis is quite difficult to break, though. Serious fractures are mainly those that have occurred in a high-energy crash or a fall from a height. As these may cause huge invisible internal bleeds due to damage to the internal organs within, injured patients are always strapped in a pelvic binder to stabilize their pelvis and prevent further bleeding until a scan has given the all-clear.

DISSECTION | *The Romans called the triangular large and heavy bone at the base of the spine (a fusion of five vertebra) os sacrum, Latin for "sacred bone." In English, too, it was called "the holy bone." The origins of "sacrum" are unclear and may stem from the ancient belief that this virtually indestructible bone, being very slow to decay, was where the soul resided.*

PELVIS & PERINEUM

GROSS ANATOMY | The pelvic organs are separated from the abdominal organs only by a thin layer of membrane; the pelvic cavity and abdominal cavities are in essence just one large space (abdominopelvic cavity). The abdominal organs, however, sit on the wide wing-like bony protrusions above the rim formed by the hip bone and sacrum (false pelvis). The area below the pelvic brim is the true pelvis. It protects the bladder, rectum and, in females, the uterus (womb), uterine tubes, and ovaries. The pelvis is separated from the genital area (perineum) by a diaphragm-like levator ani muscle (a collection of several muscles known as the pelvic floor muscles) that holds up the pelvic organs and prevents them from protruding through the openings below. The pelvic floor and muscles regulating the opening and closing of the anus below are tethered to the perineal body in front of the anus. Maintaining good muscle tone in the pelvic floor ensures urinary and fecal continence. The perineal area below the pelvic floor contains the urethra, the vagina (in women), and the anal canal. Each has an opening in the groin. The external urethral meatus is at the front and the anus is at the back. In women, the vaginal orifice is between the two, for sexual intercourse and expelling a baby during childbirth.

CLINICAL ANATOMY | The pelvic floor muscles (levator ani) form a gutter shape where they unite in the midline. During childbirth, this shape guides the baby's head into the correct orientation for exiting the uterus. These muscles stretch during pregnancy and childbirth, which can cause mild incontinence after childbirth. If the perineal body between the vagina and anus tears during a difficult labor, the results can be deeply disturbing for the woman, who may develop urinary and/or fecal incontinence. Laxity can also result in pelvic organs pushing through the pelvic floor into the perineum and genital area below. The resulting prolapse is extremely uncomfortable for the woman.

DISSECTION | *The pudendal artery, vein, and nerve are all associated with the genital area, or the perineal area. A pudendal block, for instance, is used to numb the area around the vagina before a baby's head passes through. The Latin word "pudens" means "to be ashamed," suggesting that even in ancient times the private parts were considered something to be ashamed of and hidden away.*

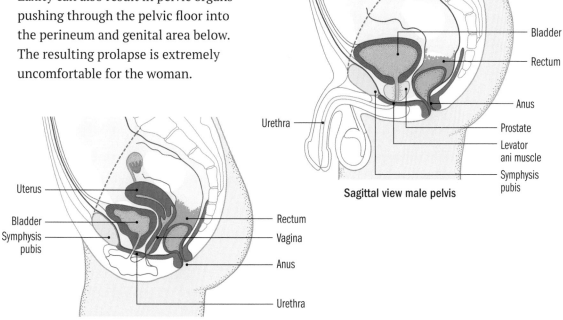

Sacrum

Bladder

Rectum

Anus

Prostate

Levator ani muscle

Symphysis pubis

Urethra

Sagittal view male pelvis

Uterus

Bladder

Symphysis pubis

Rectum

Vagina

Anus

Urethra

Sagittal view female pelvis

FEMALE REPRODUCTION

GROSS ANATOMY | The female reproductive tract has numerous functions. It secretes female sex hormones, produces and releases ova (plural of ovum, meaning "egg"), admits sperm and provides a safe place for implantation to take place and a haven for the growing baby until it is ready to be born. As a result, it has a complicated anatomical system with a feedback loop linking with the brain to maintain the delicate hormonal balance required for menstruation, pregnancy, and childbirth. It is composed of vagina, uterus, uterine tubes, and ovaries. The vagina is a thin-walled receptacle for the male penis. It leads into the uterus (womb), the entrance of which is a tight narrowing (the cervix). The uterus is thick-walled and muscular. It sits fully in the true pelvis (until pregnancy, when it expands into the abdominal region) and is held in place by numerous ligaments and a sheet of membrane overlying it (peritoneum). Fanning sideways and backward toward the ovaries are two funnel-shaped uterine tubes (commonly known as "Fallopian tubes"). Each month, when an ovum is released by the ovary, frond-like structures (fimbriae) at the end of the tubes lure it into the uterine tube to increase its chance of meeting a sperm and achieving impregnation. Breast development and the menstrual cycle are also controlled by the ovaries.

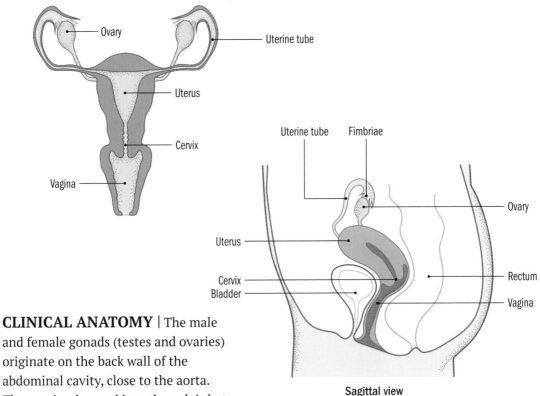

Sagittal view

CLINICAL ANATOMY | The male and female gonads (testes and ovaries) originate on the back wall of the abdominal cavity, close to the aorta. The ovaries descend into the pelvis but remain within the peritoneal cavity. A small opening allows the ovum to enter the uterine tubes. Because of its existence in a different anatomical cavity, on very rare occasions sperm can travel through the opening, and impregnation and implantation can occur outside the safe environment of the uterus. This type of ectopic pregnancy (any pregnancy that implants outside the uterus is "ectopic") can lead to rupture and severe internal bleeding, although there are some rare successful cases of abdominal pregnancies.

DISSECTION | *Until the 19th century, it was thought that the uterus (Greek, hyster) could float around the body (almost with a mind of its own) causing typically female problems. The once respectable psychiatric diagnosis of "hysteria" was attributed to this ability of the womb to cause havoc by pressing on other structures. Smelling salts under the nose and vagina, as well as sneezing, were believed to return the "wandering womb" to its proper place.*

MALE REPRODUCTION

GROSS ANATOMY | The male reproductive system lies below the bladder and between the legs. It consists of the plum-sized testes, an expandable sausage-like penis, the prostate gland, and seminal vesicles. Their combined function, after puberty, is to make and deliver sperm into the corresponding female reproductive system. The testes, lying in the scrotal sac on either side of the penis, produce sperm and male sex hormones. A long coiled tube (epididymis) attaches to each testis. It matures sperm passing through it on their way to the penis. The penis contains a very long urethra, the tube for expelling urine and semen. Covered by foreskin, a partly loose fold of tissue, it is very sensitive and made of spongy erectile tissue. During sexual arousal, the spongy tissue expands with blood, resulting in a rigid and enlarged penis (erection), whose role is to deliver sperm into a female vagina. Sperm are propelled out of the scrotum through the epididymis, along a duct (vas deferens) toward the penis. Fluids from the prostate gland and the seminal vesicles mix and dilute sperm, producing semen. This sticky, milky-white substance allows sperm to survive in the acidic female reproductive tract. Tiny bulbourethral glands secrete a lubricant into the urethra when a man becomes sexually aroused.

CLINICAL ANATOMY | Due to the long course the spermatic cord has to travel through in the scrotum and pelvis, it is vulnerable to twisting (torsion) along its course. When this happens, the blood supply to the testicles gets cut off and can lead to a permanently damaged testicle and infertility. Gently stroking the inside of the upper part of the thigh can trigger the cremasteric reflex, where the testicle on the same side lifts up due to a reflex arc involving a muscle within the scrotum. Its absence is a worrying clinical sign, which may signify twisting of the spermatic cord (testicular torsion). In young boys, the reflex is exaggerated and unreliable.

DISSECTION | *The dartos muscle is a sheet of smooth muscle enveloping the scrotum. It helps maintain the fine balance of temperature regulation required for optimal sperm production either by crinkling the scrotal skin and reducing the surface area available to prevent heat loss or by expanding the skin to cool the testicles. The testicles lift up when exposed to sudden cold due to this shriveling effect.*

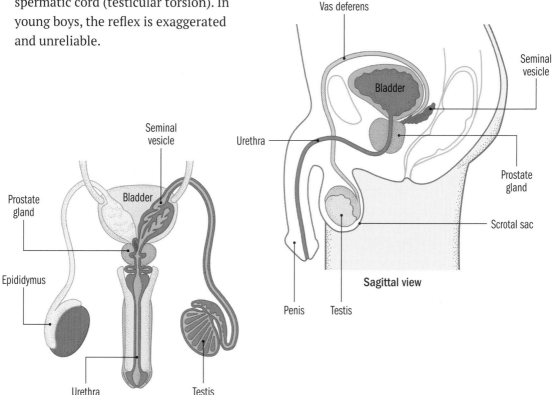

Vas deferens

Seminal vesicle

Bladder

Urethra

Prostate gland

Scrotal sac

Seminal vesicle

Bladder

Prostate gland

Epididymus

Urethra

Testis

Penis

Testis

Sagittal view

"It is the novel province of anatomy to tell the truth, the whole truth, and nothing but the truth about the structure, the origin and the history of man."

HENRY FAIRFIELD OSBORN
EVOLUTION AND RELIGION IN EDUCATION : POLEMICS OF THE FUNDAMENTALIST (1926)

4
BACK & LIMB ANATOMY

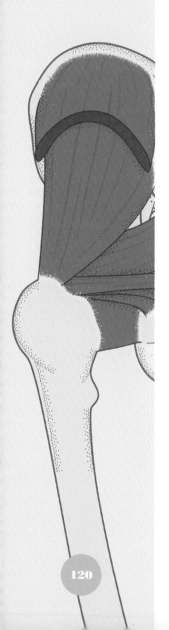

MOVEMENT

We are bipedal, and move, stand, and walk upright. In this position, our trunk is erect, our knees are almost straight, and our feet are firmly planted on the ground (unlike larger mammals who stand and move on their toes, and ungulates who use hooves at the tips of their toes). Standing up straight and moving (locomotion) require coordinated movement of upper and lower limbs, the back, and buttocks.

We sway when we stand still

Simply standing up motionless creates a center of gravity just above the hip joints and in front of the spinal column, with the center of pressure approximately midway between the insteps of our feet. The main muscles we use are two muscles in the calf (soleus and gastrocnemius), with very little activity required by our thigh muscles. As we stand, we sway slightly, a factor that deteriorates with age when we may become less able to control our posture and may fall over as a result. If you decide to stand on one foot, the weight in the upper body requires the hip to be adjusted and the muscles in the buttocks region (gluteus medius, gluteus minimus, and tensor fascia lata on either side) need to be recruited to maintain balance.

Walking is a series of repeated gait cycles

Walking looks so simple but is really a complex biomechanical process requiring synchronized movement of hips, spinal column, arms, shoulders, and head to maintain balance and to get the feet to move across the ground. A child goes through a typical pattern of learning to sit (six months), crawl (nine months), walk supported (12 months), and unsupported (18 months). By the age of three, a child's walk is similar to that of an adult.

Walking is somewhat like controlled falling: with each step, our body "vaults" itself over the limb that is not moving (inverted pendulum motion). The center of gravity fluctuates constantly, from being high in the middle of our stride to low when both our feet are on the ground. Our arms swing to stabilize our posture but, even today, how this happens is not fully understood. When we walk, we are striding through two repeating phases of the gait cycle. During one full round of the cycle, each foot is on the ground for about 60 percent of the time (stance phase) and lifted off the ground for the rest of the time (swing phase); in each of these phases, our gait goes through a single-support phase, when one foot is on the ground, and a double-support phase, when both feet are on the ground.

Three discrete things happen in the stance phase: the heel of the foot hits the ground (known as "heel strike"), then the rest of the foot hits the ground as the other foot makes contact with the ground, and, lastly, the foot leaves the ground (heel first and toes last, known as "toe-off"). Our knees are almost straight when our heels hit the ground, and bend only slightly before our toes lift off the ground. In the swing phase, two things happen: the leg lifts up and we are then propelled forward. This is how moving forward occurs, when the weight of the body is only on one limb. The knee is bent to a maximum of 60 degrees in the middle of this phase.

Muscles in the buttocks (gluteal region) tighten to stop the pelvis from dropping down toward the raised leg, as would happen if gravity took over. The gluteal, thigh, leg, and foot muscles are all in use during walking. The muscles in our calf propel us forward but this is enhanced by our foot arching and our toes flexing. The weight of the foot is balanced on the ball of the foot, the outer edges, and the heel; the toe muscles and lumbricals (small, worm-like muscles that simultaneously enable bending and straightening at three different joints in the foot) are important during walking, extending and balancing the toes so they do not buckle during toe-off.

We float a bit when we run

Running is analogous to being on a pogo stick, with our center of gravity shifting in the opposite direction to when we walk, from being low in the middle of the stride to high when we are "double-floating," when neither of our feet is on the ground. During jogging, each foot is on the ground 40 percent of the time; when we break into a sprint, this is for only 27 percent of the time. The faster you run, the less time your feet spend in the stance phase of the gait cycle. In contrast to the walking gait cycle, during running the knee of the leg supporting our weight bends. During running, too, the leg muscles exert far more forces than during walking. Our foot is moving when the heel hits the ground, and is locked into a rigid structure so that the shock of the impact can be better absorbed. Our body now vaults itself over the foot and the foot passively stretches downward, becoming more flexible to adapt to the running surface. Just before it lifts off again, it tightens and lifts upward and backward, becoming a lever. Most of us hit our heels on the ground first when running; when we sprint, to get better leverage, our forefoot hits the ground first. Curiously, our knees never fully straighten when we run.

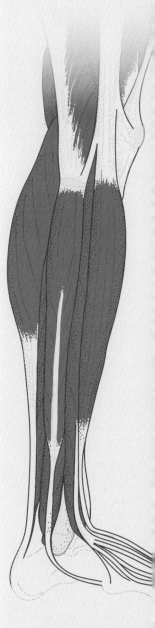

VERTEBRAL COLUMN & BACK

GROSS ANATOMY | The S-shaped vertebral column occupies a central position in the body. It provides a rigid supporting framework and protects the spinal cord running within it. While standing or walking, it transmits our weight. The curved linkage of 33 individual bones extends from atlas to coccyx (tailbone) in ever-increasing, then diminishing, size. The lowest vertebra is merely a pointy tip. Seven cervical vertebrae move the neck, twelve thoracic vertebrae help form the protective casing around the chest, five lumbar vertebrae move the lower back, five sacral vertebrae fuse to form the back wall of the pelvis, and four fused coccygeal vertebrae form the bottom of the column. There is limited flexibility between individual vertebrae but, due to cushioning intervertebral discs separating the vertebrae, the spine is very flexible. Strong back muscles, layered in the neck and lower back, attach to the spinal column, enabling movement and providing stability to allow posture. The muscles closest to the skin are the extremely broad *latissimus dorsi* and trapezius muscles. Below them are fleshy muscle bulks on either side of the spinal column known as the *erector spinae* muscles; they keep us in an upright position. Back muscles form a strong platform for limb movement and work together with abdominal wall and lower limb muscles.

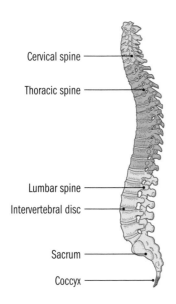

Cervical spine

Thoracic spine

Lumbar spine

Intervertebral disc

Sacrum

Coccyx

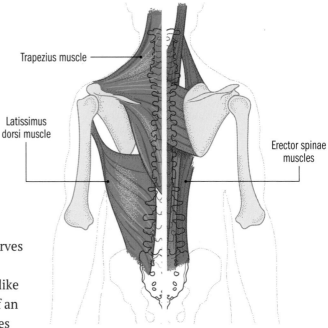

Trapezius muscle

Latissimus dorsi muscle

Erector spinae muscles

Posterior view back muscles

CLINICAL ANATOMY | Spinal nerves emerging at each vertebral level are vulnerable to damage when the jelly-like pulp (nucleus pulposus) at the core of an intervertebral disc ruptures and pushes through the tougher outer rim (annulus fibrosis). This disc prolapse presses on the nerve below and pain, numbness, tingling, and weakness (the symptoms are known as "sciatica") may radiate along the distribution of the muscles or muscle groups supplied by the nerve, usually in the area of the buttocks and/or lower limb. Most back problems, especially those in the lower back (the erector spinae and transversospinales muscle groups), can be resolved by improving core back muscle strength. Back surgery is rarely required.

DISSECTION | *On average, a female vertebral column (23½ in./60 cm) is 4 in. (10 cm) shorter than a male one (27½ in./70 cm). Intervertebral discs, which make up a third of vertebral column height, swell with fluid absorbed during periods of rest. Weight-bearing activity squeezes fluid out, the result being that a 20 percent reduction in water content and vertebral column height takes place during the day. Three hours after rising, our full height has decreased by about half an inch (15 mm).*

SHOULDER JOINT & AXILLA

GROSS ANATOMY | The upper limb is connected to the trunk via the scapula (shoulder blade) and clavicle (collar bone) meeting to form the pectoral (shoulder) girdle. They anchor the humerus (upper arm bone) to the chest via a ball-and-socket joint (glenohumeral or shoulder joint), forming the most mobile joint in the body. The scapula at the back floats over the ribcage through only muscular attachments. The ball-shaped head of the humerus nestles in a shallow cavity on the outer rim of the scapula, allowing it to freely move in multiple planes. The shoulder joint is dependent on muscles to stabilize and move it and requires several stabilizing mechanisms to prevent the head from popping out of its socket. A slice of cartilage deepens the cavity, a fluid-filled capsule surrounds it to tether the scapula to the humerus, fluid-filled sacs (bursae) pad the area outside the capsule to minimize friction, ligaments anchor bone to bone, and four small rotator cuff muscles and the biceps brachii muscle provide muscular stability. Movement is enabled by rotator cuff muscles and strong muscles (deltoid, trapezius, latissimus dorsi, serratus anterior, and pectoralis major) attaching to the shoulder complex. The axilla (armpit) is the anatomical space below the shoulder, between the arm and the trunk, filled with large vessels, nerves, and lymph nodes, all surrounded by fat.

CLINICAL ANATOMY | Mobility at the shoulder joint comes at a price: it is naturally unstable and dislocation happens relatively easily. A blow to the arm while it is raised may pop the head of the humerus forward and downward out of the socket (a ligament prevents upward dislocation) into the axilla, where nerves and vessels are at risk of damage. The deep rotator cuff muscles gripping hold of the humerus from several sites on the scapula are also prone to damage. The tendon of supraspinatus traveling within a narrow space (subacromial space) can become inflamed with overuse, and shoulder movement becomes painfully limited (impingement/painful arc syndrome).

DISSECTION | *The collarbone (clavicle) links the sternum and scapula. It is the first bone in the body to harden into bone (ossify) and the most common to become fractured. Clavicle initially meant "small key" or "bolt," perhaps because it bolts onto the shoulder. It acts as a prop to push the shoulder back and allow the free-swinging movement of the arm.*

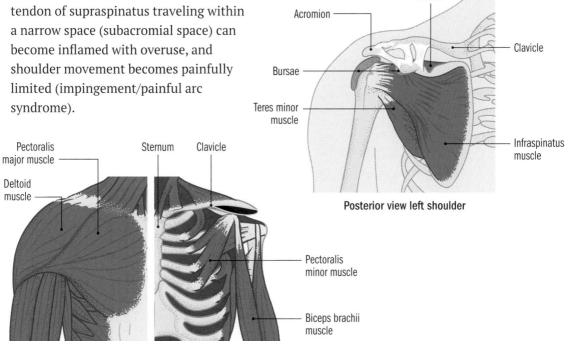

Supraspinatus muscle

Acromion

Clavicle

Bursae

Teres minor muscle

Infraspinatus muscle

Posterior view left shoulder

Pectoralis major muscle

Sternum

Clavicle

Deltoid muscle

Pectoralis minor muscle

Biceps brachii muscle

Anterior view pectoral girdle muscles

ARM

GROSS ANATOMY | The arm is the anatomical region between the shoulder and elbow joints. It consists of just one long bone (humerus) surrounded by two groups of muscles that move it (the biceps and the triceps). The humerus is the largest and longest bone in the upper limb, with a hemispherical and expanded top end (head). When the arm is resting by the side, the smooth head points slightly backward and toward the midline to form the glenohumeral joint (shoulder). A small and large bony bump (lesser and greater tubercle) can be felt close to the top of the arm. Between them, a groove carries the long head of biceps brachii tendon to provide additional stability to the shoulder joint. The humeral shaft is somewhat cylindrical, curving forward at the lower end to form two knuckle-shaped expansions that help form the elbow joint. Arm muscles are arranged into two sets separated by deep fascia. Enclosed within each fascial compartment are muscles and their blood and nerve supply. Three muscles in the biceps compartment at the front (biceps brachii, brachialis, and coracobrachialis) move the arm forward at the shoulder joint and bend it at the elbow. The single three-headed triceps muscle in the compartment at the back moves the arm backward at the shoulder and straightens the elbow.

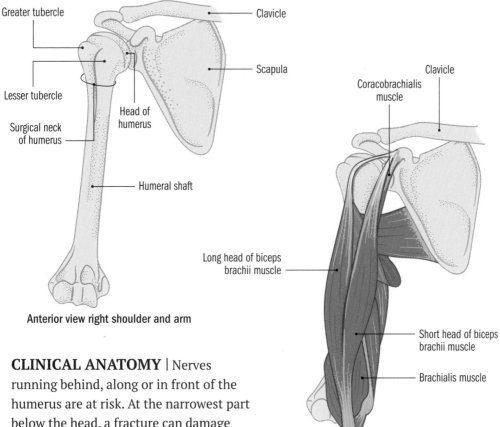

Greater tubercle

Clavicle

Lesser tubercle

Scapula

Head of humerus

Surgical neck of humerus

Clavicle

Coracobrachialis muscle

Humeral shaft

Long head of biceps brachii muscle

Anterior view right shoulder and arm

Short head of biceps brachii muscle

Brachialis muscle

Anterior view right arm muscles

CLINICAL ANATOMY | Nerves running behind, along or in front of the humerus are at risk. At the narrowest part below the head, a fracture can damage the axillary nerve running behind the surgical neck, disrupting nerve supply to the deltoid muscle. A mid-shaft fracture can damage the radial nerve running in a spiral groove in contact with the bone, leading to an inability to straighten the wrist and hand. The biceps and triceps muscles may pull the bone fragments of a facture occurring just above the elbow in opposing directions. The jagged edge of the bone protruding forward can compress the median nerve and brachial artery; a claw-like deformity may form in the hands and fingers (Volkmann's contacture).

DISSECTION | *There are several bony condyles in the body. These are usually smooth rounded bumps at the end of bones (mandible, humerus, and femur) and form joints with other bones. The Greek word kondylos originally meant "knuckle," entering the English language via Latin and French to mean "knob at the end of a bone." Anatomical condyles are somewhat knuckle-shaped.*

ELBOW & CUBITAL FOSSA

GROSS ANATOMY | The elbow is for shortening and lengthening the forearm and for reaching out and grasping objects. The lower end of the humerus forms a forward-sloping, double-knuckled expansion (the condyles) that meets the two forearm bones (the radius and ulna): the capitulum articulates with the radius on the thumb side, while the trochlea articulates with the ulna on the little finger side. They form a hinge joint that bends (flexes) and straightens (extends) the forearm, and produce a rotational movement with the radius and ulna that makes it possible to turn the palm of the hand upward (supination) or downward (pronation). The upper arm muscles act on the elbow to produce these movements (triceps straightens and brachioradialis and biceps brachii bends the elbow); only one muscle in the elbow region has a single function to bend it (brachialis). Muscles that flex and extend the forearm attach to two bony bumps above and on either side of the humeral condyles (lateral and medial epicondyles). A joint capsule and strong collateral ligaments stabilize the elbow. The cubital fossa is a triangular depression at the elbow joint within which the pulsation of the brachial artery can be felt. The median nerve supplying several muscles in the forearm and hand runs close to it. A network of veins is closest to the skin here, useful for taking blood.

CLINICAL ANATOMY | Some extensor muscles of the forearm attach to the lateral epicondyle of the humerus. Overuse of these muscles inflames their common origin, leading to pain and tenderness over the outer bump, common in tennis players (lateral epicondylitis or tennis elbow). Overuse of muscles in the flexor compartment attaching to the medial epicondyle results in symptoms on the inner aspect of the elbow; medial epicondylitis is common in golfers. The ulnar nerve runs on the inner aspect of the elbow joint, close to bone. Leaning too long on the elbows can produce numbness and tingling in the little finger and half of the ring finger, as the nerve supplies this area.

DISSECTION | *The veins at cubital fossa were commonly used for bloodletting. A slip of a barber–surgeon's knife could easily damage the underlying brachial artery and median nerve, if it were not for the "grâce à Dieu!" (grace of God) aponeurosis. Even today this flattened triangular tendon (aka bicipital aponeurosis) fans out to protect structures below the veins, providing reassurance for inexperienced clinicians taking blood samples from the area.*

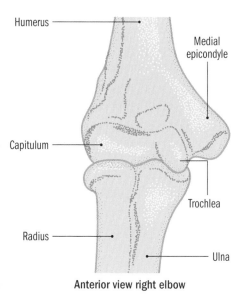

Anterior view right elbow

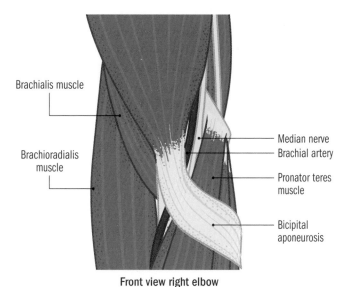

Front view right elbow

FOREARM & CARPAL TUNNEL

GROSS ANATOMY | The forearm is the area of the upper limb between the elbow and wrist. Two long bones, radius (thumb side) and ulna (little finger side), are bound together by joints at the ends of the bones, and by a thin flat sheet of ligament in the middle (interosseous membrane). This sheet divides the muscles on either side of the bones into sets: flexors at the front and extensors at the back. The 12 muscles in the extensor compartment straighten the fingers and wrist, and bend the wrist backward. The radial nerve supplies all the extensors of the upper limb. The eight muscles in the flexor compartment bend the wrist, hand, and digits (thumb and fingers) forward. The median and ulnar nerves travel with the flexors, and innervate them. The carpal tunnel is a U-shaped bony tunnel at the wrist covered by a strip of fascia (flexor retinaculum). Within it, nine tendons originating from three forearm flexor muscles (four each from flexor digitorum superficialis and profundus and one from flexor pollicis longus) tunnel their way into the hand to enable the hand's powerful gripping movements. Flexor retinaculum holds them down, preventing bowstringing of the tendons when the wrist is bent forward.

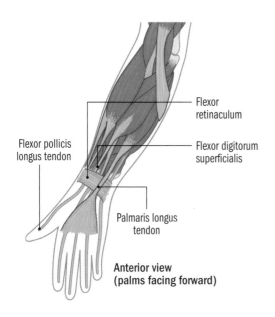

Flexor pollicis
longus tendon

Flexor
retinaculum

Flexor digitorum
superficialis

Palmaris longus
tendon

**Anterior view
(palms facing forward)**

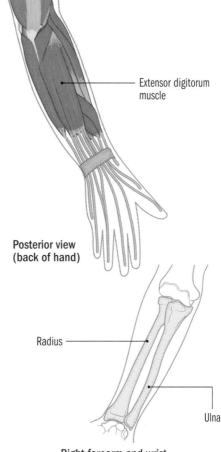

Extensor digitorum
muscle

**Posterior view
(back of hand)**

Radius

Ulna

Right forearm and wrist

CLINICAL ANATOMY | The compartments of the forearm have little space within them due to deep fascia enveloping them relatively tightly. After an injury or surgical procedure, accumulating fluid may squeeze the blood supply and nerves within so tightly that irreversible muscle death follows. Compartment syndrome is a surgical emergency. A long cut through skin and deep fascia is made to relieve the pressure. The median nerve travels through the narrow and congested carpal tunnel. If it gets squeezed tightly, pain, altered sensation, or numbness are felt within the thumb and next two and a half fingers (carpal tunnel syndrome). The flexor retinaculum can be cut through to release the pressure (carpal tunnel release).

DISSECTION | *The palmaris longus muscle may be absent on one or both sides in 10 to 15 percent of the population. This spindle-shaped flexor inserts into the wrist, anchoring the overlying skin to the underlying fascia of the hand. Absence or presence is easily tested by opposing the little finger and thumb and flexing the wrist (known as Schaeffer's test). A raised ridge running close to the midline on the front aspect of the wrist hints at its presence.*

WRIST & HAND

GROSS ANATOMY | Grasping objects and opening the palm to release them both require simultaneous contraction of numerous individual hand muscles and are mechanically extremely complex actions. The 27 bones that form the framework for these actions are organized into rows of bone: the wrist (carpus), consisting of two rows of four bones, the five metacarpals (hand bones), and the 14 bones of the five digits (phalanx). At each meeting point between the bones, ligaments and capsules tether bones to each other, allowing movement at each joint. The joint between the wrist and the metacarpal of the thumb is exceptionally mobile, enabling us to touch the tips of each finger with the thumb. Numerous muscles enter the hand from the forearm (extrinsic) and several start and end within the hand (intrinsic). Extrinsic muscles enable strong powerful movements (grasping, gripping). Intrinsic muscles fine-tune, allowing intricate movement, and are arranged into bulging groups of muscles visible on the palm of the hand—those at the thenar eminence move the thumb and those at the hypothenar eminence move the little finger. Special muscles move just the fingers (lumbricals, interossei). The digits are supremely sensitive to touch and positioning, taking up a large area in the brain disproportionate to their size to interpret these senses.

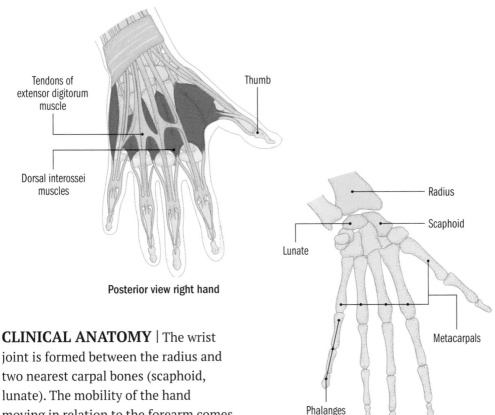

Tendons of extensor digitorum muscle

Thumb

Dorsal interossei muscles

Posterior view right hand

Radius

Scaphoid

Lunate

Metacarpals

Phalanges

Anterior view left hand

CLINICAL ANATOMY | The wrist joint is formed between the radius and two nearest carpal bones (scaphoid, lunate). The mobility of the hand moving in relation to the forearm comes at a price: the wrist is unstable and fractures often occur here. A fall onto an outstretched hand can break the end of the radius, compromising the large radial artery (felt on the thumb-side at the wrist) and the nerves in the region. The boat-shaped scaphoid is the most commonly fractured bone in the wrist but fractures are easily missed, as they show up on X-ray often only after their blood supply (which travels through a narrow isthmus of bone that usually breaks) has been compromised.

DISSECTION | *The Latin and Greek origins for phalanx (plural, phalanges) related to heavily armed soldiers in close ranks. The finger bones sit in close proximity to one another in a row, resembling a phalanx of infantry soldiers. The thumb has two phalanges, the other four digits have three each. The toe bones have a similar arrangement and are also called phalanges.*

HIP JOINT

GROSS ANATOMY | The lower limb is attached to the spinal column via the hip bone (pelvic girdle). Compared to the shoulder, this is a far more stable arrangement, as the pelvis and lower limbs must be able to bear our weight when we move around or stand still. The hip joint is a ball-and-socket joint, similar but not identical to the shoulder joint. The compromise for stability comes at a reduction in the range of movement, although the joint is still capable of a wide variety of movement. Two large bones meet to form this joint: the innominate (hip or pelvic) bone and the femur (thigh bone). The ball-like head of the femur sits snugly in a deep socket in the hip bone (at the acetabulum), held there by a strong ligament (ligamentum teres femoris) that runs from inside the acetabulum to a small dip in the head of the femur (fovea). The joint is surrounded by a fluid-filled joint capsule (resembling a purse string) that runs from the rim of the acetabulum to a raised area (intertrochanteric line) between two bony bumps (greater and lesser trochanter) close to the neck. Three tough ligaments prevent forward and backward movement of the femoral head, and fluid-filled sacs (bursae) lubricate the area. Strong gluteal and thigh muscles move the hip.

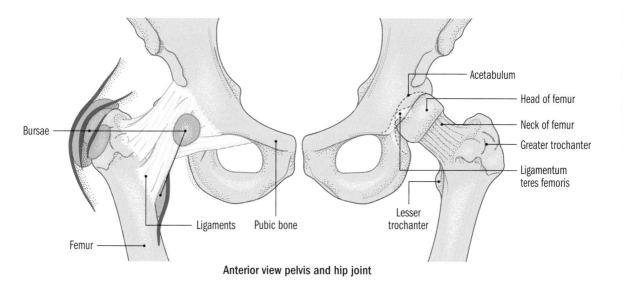

Bursae

Femur

Ligaments   Pubic bone

Acetabulum

Head of femur

Neck of femur

Greater trochanter

Ligamentum teres femoris

Lesser trochanter

Anterior view pelvis and hip joint

CLINICAL ANATOMY | Of all the joints in the human body, the hip is most important in terms of the catastrophic consequences of any damage to it. It is susceptible to age-related changes, particularly weakening of the bone (osteoporosis) and wear and tear of the joint (osteoarthritis). Weakened bone makes it susceptible to fracture at the narrow neck, particularly in the elderly. The femoral head receives its blood supply from below the joint, and blood travels up to the head plastered, surprisingly, under the fibrous capsule along the neck. So, when the neck breaks as a result of a fall, the blood supply to the head is lost and the bone dies (avascular necrosis of the femoral head).

DISSECTION | *The acetabulum is the socket into which the head of the femur fits. Its concave shape resembled a "small vinegar cup" commonly used in Roman times. The site is the fusion point of three bones forming the hip (innominate) bone. Fusion is completed between the ages of 16 and 25. In children, in whom the acetabulum (and hence the three parts forming the hip bone) is unfused, this is visible on X-ray as a Y-shape (or an inverted Mercedes Benz sign).*

GLUTEAL REGION

GROSS ANATOMY | The gluteal region is between the lower back and the upper thigh. The markedly rounded area is formed by the forward tilt of the pelvis, the fleshy mass of the largest (and, perhaps, most powerful) muscle in the body (gluteus maximus), and a considerable amount of fat. Two groups of gluteal muscles move the hip joint. The buttocks (*gluteus* is Latin for "rump" or "buttocks") allow us to sit upright without needing to rest our weight on our feet in the way that four-legged animals do. Its bony framework is the back of the pelvis and vertebral column, the hip bones on the sides and the sacrum in the midline. At the top of this region, the rounded edges of the iliac crest on the hip bone can be felt on both sides. These serve as an attachment site for gluteus maximus, which allows us to stand in an upright position by virtue of its size and the power it has to stabilize the trunk (it prevents our trunks from leaning forward). Several deep muscles (piriformis, superior gemellus, inferior gemellus, obturator internus and quadratus femoris) rotate the hip. The two rounded muscles are separated in the midline by a deep cleft (natal cleft) in which the anus sits. The bottom of the gluteal region is marked on the surface by horizontal gluteal creases, below which the thigh begins.

CLINICAL ANATOMY | Immediately below the large gluteus maximus muscle are two smaller muscles—gluteus medius and minimus. Despite their smaller size, the role they play in maintaining the trunk in an upright position when the opposite side is raised (running or walking) is vital. As long as these two muscles are functional, there is little effect on walking or running, even if the other muscles acting on the hip joint are not fully functional. Paralysis of these glutei muscles, however, results in a curious lurching gait. Nerve damage can be tested by getting the person to stand on the affected side; the pelvis on the unaffected side will dip down (known as Trendelenburg sign).

DISSECTION | *The dimples of Venus are two indentations in the lower back above the natal cleft. They are more prominent in women, in whom the dimples have been considered a sign of beauty (Venus being the Roman goddess of beauty). A short ligament tethers skin to a bony bump on the hip bone (posterior superior iliac spine), forming the dimples. They are used as surgical landmarks in spinal surgery.*

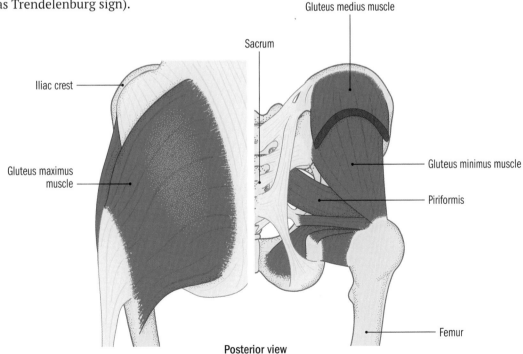

Gluteus medius muscle

Sacrum

Iliac crest

Gluteus maximus muscle

Gluteus minimus muscle

Piriformis

Femur

Posterior view

THIGH & FEMORAL TRIANGLE

GROSS ANATOMY | Some of the largest and most powerful muscles move the hip and knee and reside within the thigh. They are mostly associated with the longest, heaviest, and strongest bone in the body; the femur measures, on average, 17¾ in. (45 cm). Its ball-shaped hemispherical head contributes to the hip joint, and its lower expanded end, with double-knuckled bumps, contributes to the knee joint. Deep fascia binds together the entire lower limb, wrapping round it like plastic wrap. Where it pushes inward to attach to bone, it forms two anatomical compartments in the thigh. The muscle groups are functionally organized into three compartments, each with its own nerve and blood supply. The posterior compartment has two groups of muscles: those that move the thigh inward and upward (adductor compartment) and those that move the hip backward and bend the knee (hamstring compartment). The anterior compartment has muscles that bend the hip and straighten the knee (quadriceps compartment). When viewed from the side, the front of the thigh appears to curve forward, reflecting the shape of the curved femoral shaft overlain by the fleshy mass of quadriceps femoris (containing rectus femoris, vastus lateralis/medialis/intermedius muscles). The outer side of the thigh is flattened by a tight band of fascia, from hip bone to below the knee (iliotibial tract), which stabilizes both joints.

CLINICAL ANATOMY | A triangular space at the top of the thigh, shaped like an inverted sail, is the femoral triangle. It is bounded by two muscles (sartorius and adductor longus) and the inguinal ligament. Within it, the femoral nerve, artery and vein sit side by side (outside to inside). The pulsating femoral artery is a landmark for accessing the artery or the vein in emergencies. The femoral artery lies close to the bone lower down, as it makes its way to the back of the thigh through an opening in a muscle (adductor hiatus). Shaft fractures that tear the artery here may be catastrophic, as the artery pumps the entire blood volume through it within minutes.

DISSECTION | *The longest muscle in the body starts on a bony bump on the hip bone and travels obliquely across the thigh to insert below the knee. Sartorius (sartor is Latin for "tailor") is known as the "tailor's muscle," as tailors in former times best illustrated the use of this muscle while working cross-legged on the floor. For similar reasons, it may sometimes be referred to, rather rudely, as the "honeymoon muscle."*

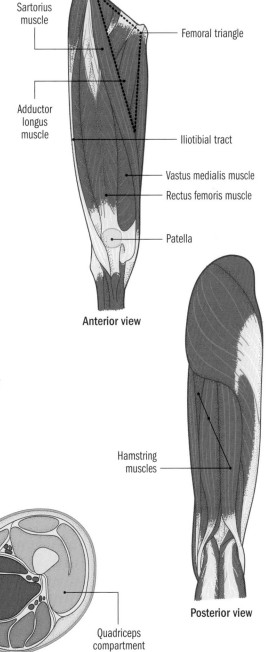

Sartorius muscle

Inguinal ligament

Femoral triangle

Adductor longus muscle

Iliotibial tract

Vastus medialis muscle

Rectus femoris muscle

Patella

Anterior view

Hamstring muscles

Posterior view

Adductor compartment

Hamstring compartment

Quadriceps compartment

Cross section of thigh compartments

KNEE & POPLITEAL FOSSA

GROSS ANATOMY | The largest joint in the body is the knee joint. It carries the entire weight of the body above it and is subject to a significant amount of relentless stress. The knee is a modified hinge joint, consisting of two joints: one between the knuckle-shaped condyles on the femur and the shinbone (tibia), and the other between the kneecap (patella) and the femur. The space between the femur and the tibia is smooth and cushioned by crescent-shaped cartilages (menisci) on the tibial plateau (widened flat top). The femur and tibia are held together by extremely strong connecting ligaments that cross over in the midline (anterior and posterior cruciate ligaments) and prevent the tibia from moving too far forward or backward when the leg is bent. Sturdy ligaments on either side (lateral and medial collateral ligaments) also hold it in place, and an incomplete fluid-filled capsule surrounds it on the sides and at the back. The quadriceps femoris tendon travels over the knee, enclosing the patella, and attaches onto the tibial tuberosity (easily felt on the upper top end of the tibia). This arrangement allows the muscles of the anterior compartment of the thigh to straighten the knee. Other thigh and leg muscles do a combination of movements at the knee.

CLINICAL ANATOMY | The popliteal fossa is the large depression behind the knee through which the blood and nerve supply to the knee and lower leg travels. The popliteal artery is the deepest structure in this space. Bulges commonly occur in the fossa and can be anything from a cyst to a weakened and ballooning artery (aneurysm). Aneurysms of the popliteal artery were fairly common when tight-fitting riding boots were used for horse-riding. If present, aneurysms require surgical repair, and care must be taken to avoid damaging the structures closer to the surface (the nerves that supply the lower leg and the popliteal vein).

DISSECTION | *The patella is the body's largest sesamoid bone. From the Latin for "sesame seed," sesamoid bones are usually very small and embedded within tendons. Their precise role is not known but they may lessen friction, and change the direction of muscle pull. Instead of patellae, babies have a soft cartilage that starts to harden into bone by their third year and is fully formed at puberty.*

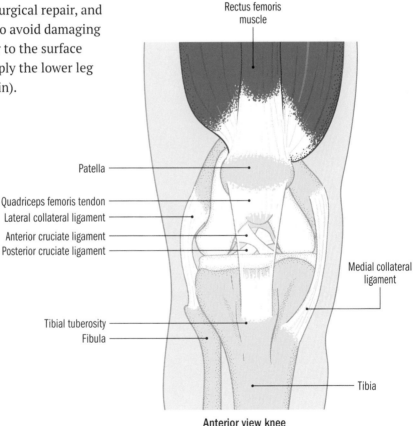

Rectus femoris muscle

Patella

Quadriceps femoris tendon
Lateral collateral ligament
Anterior cruciate ligament
Posterior cruciate ligament

Medial collateral ligament

Tibial tuberosity
Fibula

Tibia

Anterior view knee

LEG

GROSS ANATOMY | The legs (crus/crura) are anatomically situated between knee and ankle. They allow a range of movement, and enable standing and dancing. As they bear the weight of the body above, the muscles and bones within are strong. The tibia (shinbone) is the second-largest and heaviest bone in the body. The back of the leg is the calf. Its bulk and shape come mainly from a large two-headed muscle resembling a cow's belly (gastrocnemius). The front of the leg is formed by the tibia and fibula. These bones are held together by a thin flat sheet of interosseous membrane, with joints at the top and bottom. Muscles arise from both bones but mainly from the tibia. No muscles originate from the flattened front (the muscle-free zone is the shin) or lower third of the tibia. Four groups of muscles encircle the leg: the anterior crural compartment muscles pull the foot up toward the nose (dorsiflexion) and the inner foot edge upward (inversion), the posterior crural compartment muscles (in two compartments) point the foot and toes downward (plantarflexion) and bend the knee, and the lateral crural compartment muscles pull the outer edge of the foot upward (eversion). Three tendons from the posterior compartment muscles meet to form the Achilles (calcaneal) tendon that inserts into the heel bone (calcaneus).

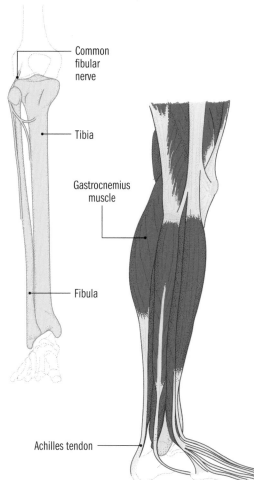

Common fibular nerve

Tibia

Gastrocnemius muscle

Fibula

Achilles tendon

Cross section

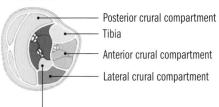

Posterior crural compartment
Tibia
Anterior crural compartment
Lateral crural compartment

Fibula

CLINICAL ANATOMY | Some structures in the leg are vulnerable to damage. The fibular (peroneal) nerve wraps around the neck of the fibula, where it can get torn in more serious knee injuries but may also be damaged when a plaster cast is placed too tightly. The nerve supplies the ankle as well as the lateral and anterior crural compartments. Loss of innervation results in a characteristic steppage gait, where the foot hangs down and the toes scrape the ground while walking (foot drop). Tibial fractures in the lower third of the bone heal poorly. As no muscles attach to this area, the blood supply is limited, meaning bone healing is slower.

DISSECTION | *The longest vein in the body is the great saphenous vein running from the big toe (hallux) to the top of the thigh along the inside of the lower limb. The origin of the word "saphenous" is shrouded in mystery. In Greek it meant "obviously visible," whereas in Semitic languages (Arabic and Hebrew) it meant "hidden." To complicate matters, it is visible at the ankle and hidden in the thigh.*

ANKLE JOINT & TARSAL TUNNEL

GROSS ANATOMY | The leg (crus) meets the foot at the ankle bone (talus). The talocrural joint (ankle joint) allows the lower limb to interact with the ground and is vital for standing, gait, and other everyday functions. It is formed where two large bumps on the ends of the tibia and fibula (medial and lateral malleoli) leave a spanner-shaped space. The talus sits like a nut within the space. The simple hinge joint allows two movements: toes to the nose dorsiflexion) or toes pointing downward (plantarflexion). The talus is wedged between the malleoli when the ankle is dorsiflexed, a very stable position. When the ankle is plantarflexed, the joint is more lax and some degree of side-to-side movement can take place. The ankle is surrounded by a fibrous capsule and is weak at the front and back. On the sides, it is reinforced by strong ligaments. On the inner aspect, the deltoid or medial collateral ligament radiates to four places on the foot from the medial malleolus. Three bands of ligaments form the lateral collateral ligament from the lateral malleolus to areas on the foot, a much weaker arrangement than on the opposite side. The tarsal tunnel is a space behind the medial malleolus covered by a band of fascia (flexor retinaculum) containing the tibial nerve, artery, and vein and muscle tendons entering the foot.

CLINICAL ANATOMY | Sprains of the ankle are very common, with the ankle being the most commonly injured joint. The arrangement of the collateral ligaments means that the individual bands forming the outer reinforcement (lateral collateral ligaments) sprain more easily. This happens when the foot is inverted too much. When the ankle is plantarflexed, it is in a compromised position, and eversion in this position can result in the tough deltoid ligament getting sprained. Although the ankle is frequently injured and bears the entire weight of the body, wear and tear of this joint (and resulting osteoarthritis) is low when compared to the hip or knee.

DISSECTION | *The talocrural joint is the only example of a true mortise joint in the body. In carpentry, a mortise and tenon joint is commonly used when a frame construction needs to be strong: the mortise is a depression cut into wood (in the ankle, a spanner-like arrangement formed by the two malleoli) and the tenon a tongue that fits into the mortise (in the ankle, the talus).*

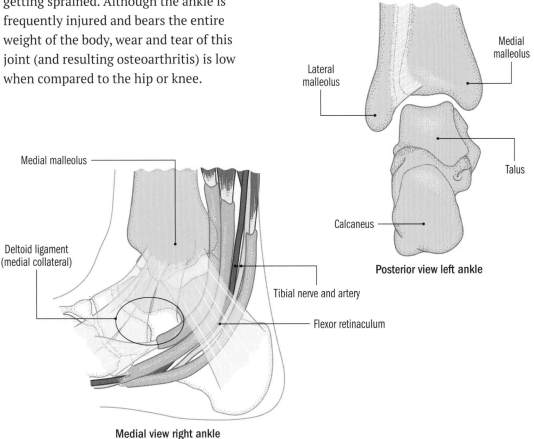

Lateral malleolus

Medial malleolus

Talus

Calcaneus

Posterior view left ankle

Medial malleolus

Deltoid ligament (medial collateral)

Tibial nerve and artery

Flexor retinaculum

Medial view right ankle

FOOT

GROSS ANATOMY | The foot is a complex and very strong functional unit with a dual role: it creates a rigid support for the body when we are stood up but transforms into a mobile springboard when we are running or walking. It has 28 bones, (including the ankle joint and small sesamoids at the base of the big toe) and over 30 joints, numerous ligaments and muscles. The foot is traditionally divided into hindfoot (calcaneus and talus), midfoot (cuboid, navicular, and three cuneiform bones), and forefoot (five metatarsals and 14 phalanges). These are joined together by muscles, ligaments, and thickened fascia on the sole of the foot (plantar fascia). The foot muscles are arranged into those entering it from the leg (extrinsic) and those starting and ending within the foot (intrinsic). Two longitudinal arches and one transverse arch, formed by the interlocking of bones, act as shock absorbers. They allow some movement. When stood up, the arches sink slightly under the load: the bones lock together and the ligaments are tensed, so the foot is a rigid base. When the foot is raised during walking, the arches unlock and allow the spring-like actions of walking and running. The heel and the metatarsal heads bear the main weight of our bodies when we are stood up.

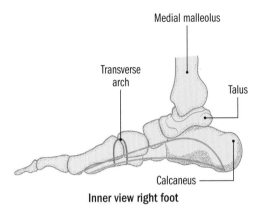

Medial malleolus

Transverse arch

Talus

Calcaneus

Inner view right foot

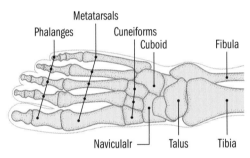

Metatarsals

Phalanges

Cuneiforms

Cuboid

Fibula

Naviculalr

Talus

Tibia

Superior view right foot

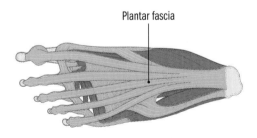

Plantar fascia

Inferior view (sole) right foot

CLINICAL ANATOMY | The plantar fascia is a thick condensed fascial band from calcaneus to the base of the toes on the sole of the foot. Heel pain associated with the plantar fascia is common. The hallmark of plantar fasciitis (inflamed plantar fascia) is pain on weight-bearing after a period of rest. Its causes are still not properly understood and recovery can take a long time. A march fracture is a type of stress fracture that occurs as a result of strain through prolonged periods of standing or repeated, concentrated damage to the metatarsals, most commonly the second or third metatarsal. Such fractures are hard to diagnose as there may be no clear fracture line on the X-ray.

DISSECTION | *Roman soldiers made the dice they used in games from sheep anklebones known as "knucklebones" (Latin taxillus or talus). We now use "talus" to indicate the second-largest bone in the foot, belonging to the group of foot bones known as the tarsal bones. Unusually, the blood supply to the talus travels past the bone then backward to supply the bone (like the scaphoid at the wrist). The bone has no muscles attaching to it.*

GLOSSARY

ALVEOLUS (plural, **ALVEOLI**)—a tiny air-filled sac within the lungs, the site at which gas exchange takes place and air from the lungs moves into the bloodstream. In the mandible, an alveolus is a tooth socket.

ANASTOMOSIS—a coming together of two completely separate blood vessels that ensures areas in the body are covered by blood arriving from multiple sources.

ANEURYSM—a weakened area in a blood vessel that bulges out and looks like a balloon. Aneurysms can occur anywhere in the body but can be particularly catastrophic if they burst either in the brain (brain aneurysm) or in the aorta (aortic aneurysm).

ANESTHESIA—in clinical practice, this refers to artificially numbing part of the body to pain, usually with intravenous drugs but sometimes also with drugs injected into the space around the spinal cord, or into the skin for a more localized effect.

APONEUROSIS—a flattened sheet of tendon connecting muscles.

AXON—the extension of a nerve cell (neuron) along which impulses travel from the cell body of one cell to other cells. They are long and wiry.

BILE—a yellow-green-brown fluid produced by the liver that breaks down fatty substances in food. It is stored and concentrated in the gallbladder.

BILIRUBIN—a yellow-orange substance formed by the breaking down of blood cells (hemoglobin). Accumulation of it in the blood and tissues gives the yellowy appearance of jaundice in the skin and sclera of the eye.

CAPILLARY—hair-thin (one-cell-thick) branching blood vessel that unites with other vessels to form a network between arterioles (medium-caliber oxygenated-blood-carrying vessels) and venules (medium-caliber deoxygenated-blood-carrying vessels). Gas exchange occurs in capillary beds.

CARTILAGE—a firm but flexible material found in several structures (external ear, larynx, respiratory tract, joints) giving them shape and the ability to move slightly. Most of the skeleton is cartilage in embryonic and fetal life but becomes bony at various stages in growth.

CELL—microscopic structures in the body, the smallest and most basic unit of living matter in an organism. Each cell is surrounded by a cell membrane, and contains genes, a nucleus, a fluid that enables chemical reactions (cytoplasm), and organelles (small structures with specialized functions).

CILIA (singular, **CILIUM**)—minuscule hair-like projections on cells that can move particles along. In the respiratory tract, they sweep together and push away fluid and particles.

CEREBROSPINAL FLUID (abbreviated to **CSF**)—clear fluid bathing the brain and spinal cord, acting as a shock absorber, lightening the weight of the brain and supplying it with nutrients. CSF is continuously produced and absorbed by the body.

CERVICAL (Latin, **CERVIX** means neck)—anything referring to the neck, either in the neck region between the head and trunk (as in cervical vertebra) or the cervix of the uterus (the narrow neck of the uterus at the upper end of the vagina that opens and widens during childbirth).

COMPARTMENT—anatomical or fascial compartments within the body that are surrounded by fascia and contain a group of muscles with a functionally similar action (for instance, biceps and triceps in the arm) as well as their blood and nerve supply.

CONNECTIVE TISSUE—tissues in the body that bind, support, separate, or connect structures (for instance, bones, ligaments, tendons, blood vessels, and cartilage). They have very few cells and are embedded in a matrix without cells (extracellular).

DIABETES—a common disease where the body is less able to respond to the hormone insulin or produce this hormone, resulting in high levels of glucose in the blood (normally glucose is mopped up from the blood by insulin). Two main types, Type 1 and Type 2.

DIALYSIS—a machine used to purify blood in a similar way to the kidneys and used in patients who have kidney damage. Prevents toxins and waste products from accumulating in the bloodstream.

DEOXYRIBONUCLEIC ACID (abbreviated to **DNA**)—present in most organisms, a self-replicating material. In living cells, DNA is found in chromosomes and contains our genetic instructions when linked together (gene).

DUCT—a tube that normally carries a secretion from a gland (for instance, the parotid duct carries saliva from the parotid gland into the mouth).

EMBRYOLOGICAL—refers to the earliest period within the development of the unborn individual. An embryo starts its life at fertilization (when an ovum and a sperm unite) until it is eight weeks old. After this stage, the embryo becomes a fetus.

ENDOCRINE—refers to glands that secrete hormones straight into the blood. Exocrine glands secrete their products via ducts (and not directly into the blood).

FASCIA (plural, **FASCIAE**)—a thin sheath (or sometimes layers) of fibrous tissue that surrounds muscles, organs, and vessels.

FISSURE—usually a naturally occurring groove or cleft in an organ or between body parts. Fissures are found on the brain and in the lungs.

FETUS—refers to the growing unborn human from week 8 after conception until birth. This is the stage when distinctly human features start to take shape.

FORAMEN—(plural, **FORAMINA**) any naturally-occurring openings or passages in the body, usually through a bone or into a bone.

FOSSA (plural, **FOSSAE**)—any shallow pit, cavity, or depression in a bone.

FRACTURE—a break in the continuity of a bone or cartilage (or the like). These usually need to be realigned so healing can occur.

FUNDOSCOPE (also known as **OPHTHALMOSCOPE**)—used to inspect the inside of the eye (fundus of the eye). Best utilized in a dark room when an eye drop has widened the pupil.

GANGLION (plural, **GANGLIA**)—a swelling on a nerve fiber with lots of nerve cell bodies within it, occurring outside the central nervous system.

GENITALIA—referring to the organs of reproduction (genitals) but mainly to those occurring externally.

GLAND—a group of cells that are specialized in function to do a particular job. Their main purpose is to produce chemical substances. Exocrine glands secrete their products via ducts (and not directly into the blood), while endocrine glands release products directly into the bloodstream for transportation to a distant site.

HERNIA—organs or fatty tissue can squeeze through weak areas in muscle walls or fascia. If they cannot go back to where they were intended to be, they may get trapped and the blood supply cut off.

HORMONE—chemical substance that regulates the way the body works, functioning like a complex signaling service. Normally, hormones are produced in specialized cells in glands and transported via the bloodstream to an organ or part of the body at a distant site.

INSERTION—muscles insert and originate at different places. The insertion of a muscle is where it is anchored on the bone, usually further away from the torso. When a muscle contracts, the insertion of the muscle is what moves the structure it is attached to.

JOINT—bones meet at junctions throughout the skeleton and are fitted together in a number of ways, some allowing movement and some not allowing movement. They are generally grouped together based on the type of movement allowed at the joint.

LACRIMAL—relates to the gland on the outer edge of the upper eyelid, which secretes a watery substance that helps to form tears and the tear film covering the eye. Lacrimation is the flow of tears.

LIGAMENT—bones are held together by tough but flexible bands of fibrous connective tissue and are similar to tendons. Although flexible, they do not stretch. Sometimes ligament also indicates structures that support or connect organs (there are several holding the uterus in place and others on the liver, for instance).

MAMMARY (Latin, **MAMMA**, probably a child's first word for mother)—relates to breasts in females. These are milk-secreting glands from which babies receive their nourishment in infancy.

MASTICATORY—refers to the process of chewing and grinding food by the teeth and using muscles that do this in the head region, the masticatory muscles (there are four).

MATRIX (plural, **MATRICES**)—cells of connective tissue are embedded within a meshwork of material outside the cells (extracellular). Matrices are very variable in consistency—fluid in blood, firm and tough in cartilage, and hard in bone.

MEMBRANE—in gross anatomy, membranes are very thin pliable sheets of tissue that cover surfaces or enclose structures. The lungs, heart, brain, and some internal organs are enclosed within layers of membrane (pleura, pericardium, meninges, and peritoneum).

MIMETIC—mimetic muscles are a group of muscles in the face that are used to express emotions (for example, smiling or frowning). Also known as facial muscles or muscles of facial expression.

NERVE—white fibrous and wiry structures that transmit signals to the brain from muscles or organs. Nerves are enclosed in a cable-like sheath that protects the axons within. Afferent nerves carry sensation (touch, sound, etc.) to the central nervous system, and efferent nerves convey signals from the central nervous system to muscles or glands or organs. Some have a mixed function.

ORGAN—a self-contained part of an organism which usually has a specific function. Organs are formed by a group of tissues coming together to do a specific job or jobs.

ORIGIN—the place where a muscle starts on a bone, usually closer to the trunk than the peripheries, and this is a fixed attachment on a bone. The muscle has more bulk at its origin than at its insertion (where it attaches away from the torso).

ORGANISM—a form of life that is considered an entity but consists of interdependent parts that work together to sustain life. Humans are organisms (as are animals, plants, fungi, viruses, etc.). Organisms grow, reproduce, react to stimuli, and maintain a balance in the body (homeostasis).

PERINEAL—relating to the perineum, a diamond-shaped area below the pelvic diaphragm (a sheet of muscles separating pelvis and perineum) and between the thighs (between the vulva and the anus in women and between the scrotum and anus in men). It is an erogenous zone in both sexes.

PLAQUE—in medicine, these refer to clumps of cholesterol and other substances that stick on the inside of vessels and narrow the available space for blood to flow through. If they block the vessel, the area supplied by the artery becomes starved of blood. This is what happens in a heart attack or stroke.

PLASMA—clear and colorless fluid, essentially blood without platelets or red and white cells.

PLEXUS—an intricate web-like network of blood vessels, lymphatic vessels, or nerves in the body, the arrangement of which is usually complex. For instance, the brachial plexus is a bundle of nerves beginning at the base of the neck but becoming a tangle of closely connected nerves, with overlapping regions, and extending into the armpit to supply the entire upper limb.

POTENTIAL SPACE—these are spaces in the body between two structures that, normally, are closely pressed together, and are not real physical spaces. They have the potential to become spaces in disease, when they are filled with fluid or air, such as when air fills the space between two tightly pressed-together layers surrounding the lungs.

PREMATURE—a premature baby is born before 37 weeks of gestation, sometimes much earlier, which can be associated with risks of complications. The normal duration of gestation in humans is around 40 weeks, although delivery anywhere between 37 and 42 weeks is considered normal.

PROLAPSE—when an organ slips out of its normal place due to weakening of the supporting structures. Commonly occurs with the uterus, vagina, and bladder, but other organs can also prolapse.

REFLEX—involuntary actions by the nervous system to a stimulus. They cannot be controlled or overridden by conscious thought.

RETINA—the light-sensitive area at the back of the eye. Light is focused onto the retina via the lens, and impulses created on the retina are sent via the optic nerve for processing at the back of the brain (in the occipital lobe).

SINUS—usually a cavity within a bone, especially within the bones of the face. The paranasal sinuses in the facial skeleton are air-filled cavities lined with respiratory mucosa.

SPHINCTER—a ring of muscle that opens and closes tightly to allow substances to pass through at a desired pace or to guard an area. They are numerous in the body.

TENDON—cordlike band of fibrous tissue that connects a muscle to a bone (mainly). They are tough and do not stretch.

TISSUE—a group of cells with a similar structure that are woven together in an extracellular matrix to work together for a specific task. Organs are comprised of many groups of tissues working together.

FURTHER READING

Brassett, Cecilia, Emily Evans, and Isla Fay. *The Secret Language of Anatomy*. London: Anatomy Boutique Books, 2017.

Delaney, Conor P. *Netter's Surgical Anatomy and Approaches*. Philadelphia: Saunders Elsevier, 2014.

Ellis, Harold and Vishy Mahadevan. *Clinical Anatomy: Applied Anatomy for Students and Junior Doctors*. Chichester: John Wiley, 2013.

Moses, Kenneth P., John C. Banks, Pedro B. Nava, and Darrell K. Petersen. *Atlas of Clinical Gross Anatomy*. Philadelphia: Saunders Elsevier, 2012.

Norton, Neil S. *Netter's Head and Neck Anatomy for Dentistry*. Philadelphia: Saunders Elsevier, 2012.

Roberts, Alice. *Human Anatomy: The Definitive Visual Guide*. London: Dorling Kindersley, 2014.

Cunningham, Daniel J., and George J. Romanes. *Cunningham's Manual of Practical Anatomy*. Oxford: Oxford University Press, 2016.

Sinnatamby, Chummy S. *Last's Anatomy*. New York: Churchill Livingstone Elsevier, 2011.

Standring, Susan. *Gray's Anatomy: the Anatomical Basis of Clinical Practice*. New York: Churchill Livingstone Elsevier, 2016.

INDEX

ABOUT THE AUTHOR

Dr Joanna Matthan *MA (English), MBBS, PGDipClinEd, SFHEA* is Lecturer at the Faculty of Medical Sciences, Newcastle University, England, with a background in Medicine and English and a previous career in the corporate world. She sidestepped from clinical medicine into medical education and now predominantly teaches anatomy to medical and dental students. She also teaches clinically relevant anatomy to postgraduate medical graduates training to be surgeons, anesthetists, radiologists, and other healthcare professionals, as well as promoting anatomical knowledge widely through public engagement. Her research interests range from anatomy pedagogy to those in the wider sphere of clinical education. She is Councillor of the Anatomical Society, and a member of the British Association of Clinical Anatomists.

Picture credits

ACKNOWLEDGMENTS

An anatomy book can never be solely one's own achievement. Countless pioneering doctors, anatomists, and artists have collated knowledge on the human body. I continue to be astounded by their arduous and indefatigable spirit, which has set the foundation for all of our clinical knowledge. Innumerable body donors have gifted their earthly vessels for us to learn and teach from; their generosity of spirit enables us to continue expanding our shared knowledge. My debt to them is immeasurable. The medical and dental students with whom I have spent countless hours poring over the marvels of the human body in the dissecting room constantly inspire me to do a better job and learn more—and make me laugh on a daily basis. Despite the somber setting, learning and teaching is never as much fun anywhere else. I hope I can inspire a few of them to learn more anatomy than is needed to pass an exam.

I owe a debt of gratitude to several of my academic colleagues and educators. Dr Roger Searle made learning anatomy in the dissecting room both intellectually stimulating and enormous fun during my foundational years as a medical student, and took a chance on me by giving me my first job in academia. Professor Stephen McHanwell instilled in me a deep love for the head and neck region, ever since my first masterclass in his presence. Through her generous spirit, Dr Gabrielle Finn has opened doors for me that would have remained firmly shut, and I am indebted to her in countless ways. Just a few minutes in the presence of Dr Pauline Prabahar taught me how very little I knew about the human body, and how much more there is to learn that can be applied to clinical care.

A chance encounter at a dinner led to an enduring relationship with Professor Jan Illing, a mentor like no other and a paragon of professionalism and compassion for whose presence I am truly grateful in my life. Dr Laura Delgaty is a phenomenal role model straddling three domains in my life, that of esteemed colleague, exceptional educator, and trusted friend. My three children—Samuel, Daniel, and Maria—are my greatest critics and truest friends, and keep me grounded. Without them in my life, I would achieve little and aspire to nothing. Life is meaningful—and anatomy, too, comes alive—because of them.